# ISLAND IN THE SKY

## Building the International Space Station

Piers Bizony

**AURUM PRESS**

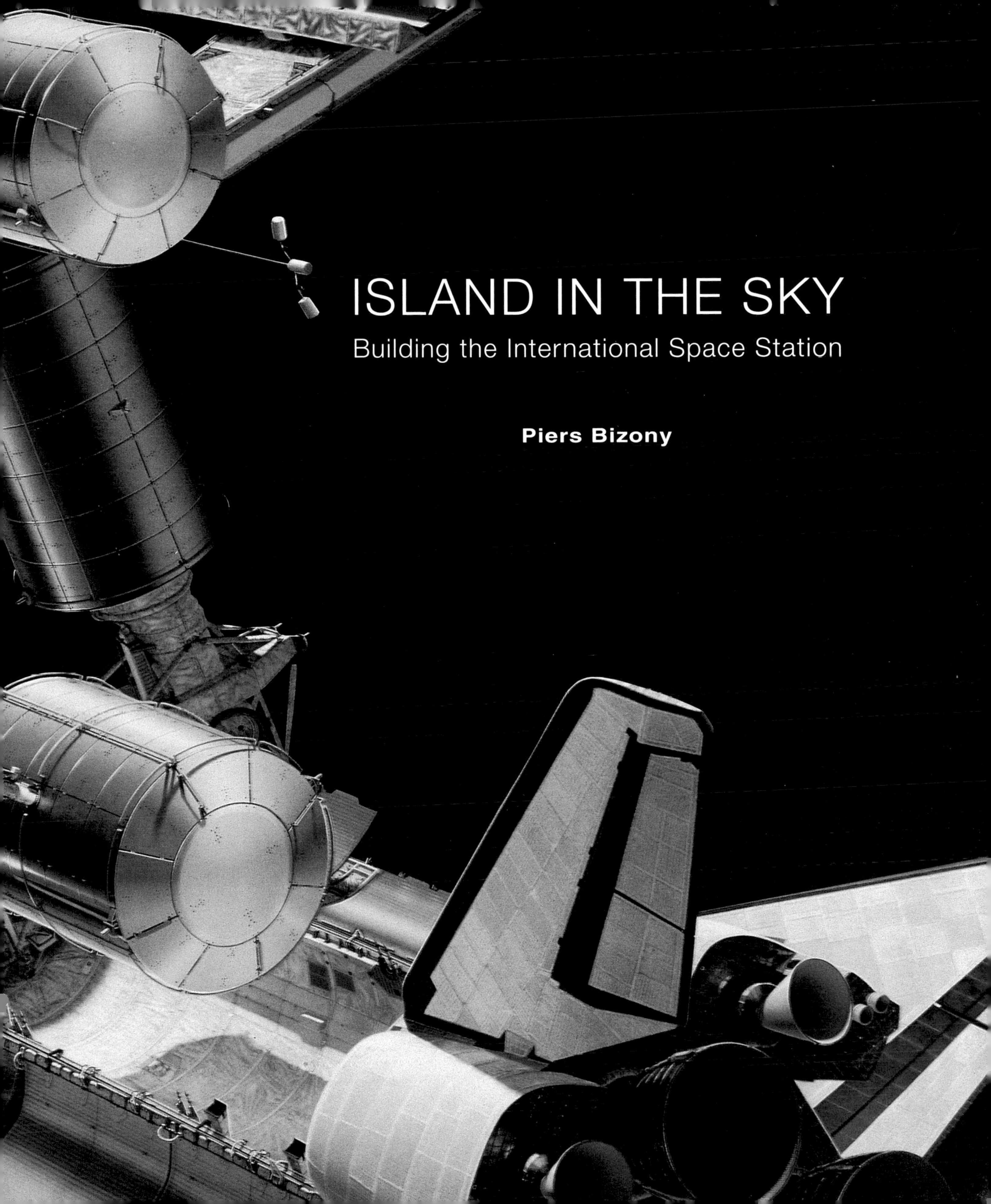

# ISLAND IN THE SKY

## Building the International Space Station

**Piers Bizony**

First published in Great Britain 1996
by Aurum Press Ltd, 25 Bedford Avenue, London WC1B 3AT


A catalogue record for this book is available from the British Library

ISBN 1 85410 436 5

2 4 6 8 10 9 7 5 3 1
1997 1999 2000 1998 1996

Designed by Piers Bizony

Printed and bound in Hong Kong
by Paramount

# CONTENTS

# FOREWORD

We are about to embark on the greatest and most expensive space project since the first Apollo moon landing in 1969. It will be the largest international scientific and technological collaboration in peacetime history. America and Russia, for so long the main opponents in a global Cold War, will work very closely together, along with Canada, Japan, Germany, France, Italy and seven other European countries. This project is the International Space Station. It will cost over $20,000,000,000 (20 billion) to build, and perhaps as much again to service and operate over its 15-year lifespan.

The station was first conceived by American engineers who wanted to use it as an orbiting garage for lunar and interplanetary spacecraft, but in recent years its primary mission has been redirected towards activities that have some prospect of yielding more immediate practical benefits: in particular, medical research in the weightless environment, which may lead to better drugs and treatments. In 1984, President Ronald Reagan gave his approval for a space station. For the last decade, American politicians have argued bitterly about its merits. Is access to weightlessness really *worth* $20 billion, or can the money be better spent in other ways? Maybe the project's true value lies not so much in its capacity as a research platform, but as a tool of geopolitics? The station's scientific justifications have become entangled with diplomatic and commercial factors that have less to do with what happens in space, and rather more to do with events down here on the ground.

But the space station *is* happening. All the paper plans have been transformed into solid hardware. A myriad of modules, laboratory compartments, solar cell arrays, docking rings and other equipment is nearing completion in factories scattered around the globe. The first component, a Russian-built propulsion and power module, is due for launch in late November 1997. By May 1998 a Russian service module with a small temporary living and work compartment will have been added, along with American docking adapters and solar cells. Over the next four years America will dock two large modules, one for scientific reasearch and the other to serve as a permanent living compartment. Europe and Japan will provide two more scientific lab modules between them, along with equipment storage canisters and unpressurized experiments platforms.

The final structure, with its long latticework metal trusses and sprawling solar power arrays, will weigh 460 tons and span an area equivalent to two football fields. Its interior living and working space, a sprawling warren of labs, sleep compartments, washrooms, cupolas, tunnels and airlocks, will amount to the cabin space of two Jumbo Jet airliners. In all, 30 major rocket flights will be required to bring the space station into operation, using at least four different kinds of launch vehicle, manned and unmanned, fired from three continents. One of them, a European booster, will be a brand-new design with the very latest rocket engines, more advanced even than the propulsion systems used by America's space agency, NASA. At the other end of the scale, Russia will use missile-derived boosters and cramped crew capsules dating from the earliest days of space exploration.

The space station promises to be an incredible machine, complex, ambitious, awe-inspiring, exactly the kind of project we might *expect* to see at the dawn of the New Millennium. But there are great risks to overcome, and the entire project is haunted by disturbing contradictions.

*Island in the Sky* explores the station's origins, not just as a space endeavour but also as a product of political and industrial forces. The story is told mainly from the American point of view, and in particular from the perspective of NASA, who first proposed that the station should be built. But this is not to suggest that the space agencies of Europe, Canada and Japan are any less crucial to the station's development. As for Russia, their engineers are considerably more experienced than America's when it comes to building space stations, and their important programs are discussed in this book.

## THANKS TO THE FOLLOWING:

I am indebted to dozens of people within NASA and other organizations for their many insights into the space station program, and for providing substantial documentary and photographic material. Connie Moore in the Photographic Records Office at NASA's Washington headquarters allowed me full access to thousands of picture files, and then processed my requests with great speed. Dr Roger Launius at the NASA History Office in Washington provided valuable documentation, most of it from sources entirely independent of NASA itself. Dr Michael Wright provided an equivalent support from the History Office of the Marshall Spaceflight Center in Hunstville, Alabama. In Washington, Ray Castillo briefed me thoroughly on space station progress. At the Johnson Space Center in Houston, Mike Gentry provided me with a mass of photos. Fred Ordway put me in touch with Huntsville personnel. Rose Gottemoeller at the International Institute for Strategic Studies in London provided me with key insights and documentation about relations between America and Russia. Tim Furniss, space editor of *Flight International* magazine answered many questions for me. As for the contractor companies who are actually building the space station, the public relations staff at Boeing, McDonnell Douglas and Rockwell were tirelessly helpful.

I am grateful to Simon Atkinson for providing some superb pencil drawings for this book. My thanks also to David Hardy for the use of a superb artwork. The Novosti Press Agency in London, and the Science Photo Library provided key photos of non-American space hardware.

**A WHEEL-SHAPED SPACE STATION**
In this 1952 painting by Chesley Bonestell, a wheel-shaped space station is attended by a delta-winged shuttle rocket. The station gently rotates to provide its crew with artificial gravity.

**INSIDE THE WHEEL**
Rolf Klep's 1952 painting shows the scale of the rocket planner's ambitions.

# THE LAST FRONTIER

*In the 1950s, rocket experts dreamed of founding a busy infrastructure in earth orbit, a 'staging post' where space crews could build giant ships destined for the moon and beyond. The key component would be a large rotating space station attended by winged shuttle rockets. Versions of all these ideas appeared in books and popular magazines. After 1961 a lunar landing became a national priority, but a space station was no longer needed.*

# THE LAST FRONTIER

***'Scientists and engineers now know how to build a station in space that would circle the earth 1,075 miles up. The job would take 10 years, and cost twice as much as the atom bomb. If we do it, we can preserve the peace and take a long step toward uniting mankind.'***

Such a quote might easily have been sampled from the latest NASA space station press release, yet this particular vision was presented to a mass audience more than forty years ago in a series of articles for *Collier's*, a popular magazine of the time. Between 1952 and 1954 seven major space articles were published, including descriptions of a lunar colony and a mission to Mars. These ideas had been familiar to rocket experts and science fiction enthusiasts since the 1930s, but the *Collier's* articles represented perhaps the first time the general public had been asked seriously to consider rocket ships and space stations and trips to the moon as elements of a national policy.

*Collier's* regularly sold 3 million copies a month. As a family title, it would have been read by perhaps 15 million people. The space articles stimulated a succession of popular books and television shows (including an influential space documentary from Walt Disney), and undoubtedly they helped prepare American taxpayers to accept the forthcoming expenses of the Space Age. The first feature of March 22, 1952 was entitled 'Crossing the Last Frontier'.

In *Collier's*, the German-born rocket engineer Wernher von Braun outlined his most cherished ambitions for a peaceful use of rocket technology. His concepts were brought vividly to life by illustrators Chesley Bonestell, Rolf Klep and Fred Freeman. Von Braun imagined a wheel-shaped space station gently turning on its axis, with its crew enjoying the artificial gravity generated by the rotation. A famous painting by Bonestell shows the station attended by winged rocket planes, while in the foreground, three huge landing ships are prepared for their trip to the moon 'within the next 25 years'.

Such concepts owed their origins not just to rocket theory but also to America's romantic self-image of its origins—Columbus, the *Mayflower*, the pioneer trail. A small fleet of ships sets sail across the ocean to discover new territory; a settlement is established at the point of landing; the settlement becomes a town; the inland areas yield riches, and a busy

trade route is established with the fleet's home port. In time, other ships arrive carrying families of settlers to populate the new territories. Right up to the late 1970s, theorists continued to talk in terms of 'conquering' and then 'colonizing' the space 'frontier'. Always, their plans were worked out with exact attention to fuel weights, rocket thrusts, orbital heights and speeds. As the renowned futurologist and space writer Arthur C. Clarke has pointed out, 'No achievement in human affairs was ever so well documented *before* the fact as space travel.'

**SCREEN STATION**
In 1968 Stanley Kubrick and Arthur Clarke's science fiction film *2001: A Space Odyssey* portrayed a giant orbiting hotel tended by winged space liners. The wheel shape is derived from the *Collier's* concepts. This sort of image still springs to mind when we think of the phrase 'space station', but the real thing, when it is finally built, will be very much smaller and less luxurious.

## THE PROFIT PROBLEM

The flaw in the plan was simple: where was the trade? Ships could be sent into space, but what could they be expected to bring back as a return on the huge investment of launching them? This fundamental economic problem could not be wished away by any amount of romantic language. *Collier's* demonstrated even to the most sceptical audience that space flight was possible, and that there were many good scientific and spiritual reasons for initiating some kind of program; but there seemed no obvious way to justify spending the vast sums involved in building a huge space station or the fleet of moon ferries, let alone ships bound for Mars. In the event, politics rather than commerce sent humanity into space. Von Braun might have been surprised to know that rockets would reach the moon eight years sooner than his *Collier's* deadline.

## FIRST IN SPACE

The first real-life steps into the cosmos turned out to be very different from anything that *Collier's* readers could ever have imagined—or even dreaded. On October 4, 1957, Soviet Russia launched 'Sputnik,' the world's first satellite. The conquest of space became rooted not in trade or economics, nor even in scientific return, but in simple political posturing. The economic arguments for building rockets became strangely inverted. By accepting the huge expense of these abstract adventures without prospect of a real financial return, Russia and America set out to show what their rival economies and societies might be capable of down on earth.

Popular and press reaction in America to the launching of Sputnik was intense and largely fearful. President Eisenhower couldn't understand why such a tiny and harmless satellite was causing so much fuss. He had little interest in space activities, and was convinced that this 'foolish' Soviet experiment would fizzle out after a while. But he was wrong.

Eisenhower's most consistent critic on Capitol Hill, at least on space issues, was Senator Lyndon Johnson, who made a series of influential speeches in support of an expanded rocket program. 'First in space means first in everything', he urged. In 1958 America's disorganized space projects were pulled together under a civilian authority, the National Aeronautics and Space Administration (NASA). Much of the new agency's strength stemmed from Johnson's legislative skills.

A year later, President Eisenhower reluctantly approved a new rocket called 'Saturn', to be designed by Wernher von Braun's team. Contrary to

**HOW IT COULD HAVE BEEN** A Chesley Bonestell painting for *Collier's* depicts three lunar ferries nearing completion in earth orbit, with the space station and ferry shuttle in the distant background. During the 1950s, Wernher von Braun and his colleagues were convinced that the moon could not be reached directly from earth, because any booster loaded with enough fuel for the trip would be too heavy to get off the ground. Large lunar landing ships would have to be constructed in orbit above the earth, using smaller rockets to ferry up components from the ground. The builders of these ships would live in 'space stations', like pioneers in a new frontier town. The landing ships would be powerful enough to ferry components for a large and permanent scientific base across and down to the lunar surface. Meanwhile, the space station would serve as a communications relay and navigation beacon for people on earth.

the view of popular history, the early Saturn program was not specifically focused upon a moon voyage. Within NASA, the belief was growing that a lunar landing might indeed be possible, but the agency's chiefs decided to exercise political discretion. In December 1959, at the end of his eight-year reign in the White House, President Eisenhower made it clear he would never sanction such a project.

All this caution was negated by the rocket activities of Soviet Russia. NASA had caught up as far as unmanned satellites were concerned, but on April 12, 1961 the Soviets beat America yet again, launching 'cosmonaut' Yuri Gagarin into orbit in a spherical silver-clad capsule called 'Vostok'. And this was not the only worry for America's new

Democrat president, John F. Kennedy. Less than a week after Gagarin's flight, a CIA-backed mercenary force was cut to pieces while attempting to create insurgency within Fidel Castro's Cuba.

Kennedy was anxious to divert attention from the Cuban fiasco, and to restore national prestige. On April 20 he sent a memo to Lyndon Johnson, now Vice-President, asking him to consider ways in which the US could achieve a lead in space: 'Do we have a chance of beating the Soviets by putting a laboratory in space, or by a trip around the moon, or by a rocket to land on the moon? Is there any other space program which promises dramatic results in which we could win?'

Three weeks later, on May 5, NASA launched Alan Shepard into space, albeit not into a full orbit. Shepard rode for 15 minutes inside a tiny capsule, the 'Mercury', hauled aloft on the tip of a converted Redstone intercontinental missile. It was a crude flight, but it was enough to restore American self-esteem. NASA's engineers had delivered on their promises, and in a commendably short time of less than three years. But Mercury wasn't enough. America hadn't been *first*.

Kennedy was encouraged by the success of Alan Shepard's flight. He examined the possibilities for future space programs as reported to him by Lyndon Johnson. The 'laboratory in space' option was realistic, affordable and scientifically useful, but there was every possibility that Soviet engineers might build one themselves; and for Kennedy it didn't seem to promise the required 'dramatic results'. On May 25, 1961 he announced his decision on space policy in a famous speech before Congress: 'I believe that this nation should commit itself to achieving the goal, before this decade is out, of landing a man on the moon and returning him safely to earth.'

**SATURN'S FATHER**
Wernher von Braun built the V2 missile for Nazi Germany. In 1945 he and many of his team surrendered to American troops. From the late 1950s von Braun was allowed to work on civilian American space programs, and he became chief of NASA's Marshall Spaceflight center, spearheading construction of the Saturn rocket that carried Apollo to the moon.

## THE APOLLO PROJECT

Wernher von Braun knew that a more powerful version of Saturn could probably reach the moon, but nobody yet understood how to make any part of it touch down safely. The lunar vehicle would have to land on its stern, and later, take off again without the benefit of a launch gantry and ground crew. It would also have to carry heavy fuel and equipment for the return voyage—a separate problem in itself.

During 1961, John Houbolt and a small group of colleagues within the space industry conceived the basis for a landing system built from ultralight components. There would be no need for multiple launches, nor for building anything in earth orbit. The basic methodology was known as Lunar Orbit Rendezvous (LOR). The beauty of the LOR concept was that it incorporated its own 'staging post'. The heavy reentry capsule and all its systems for the return trip to earth would be left in orbit above the moon, while a separate lander went down to the surface. Then the lander's top stage—a structure so lightweight and fragile you could punch a hole in its skin with a screwdriver—would come up and rendezvous with the return ship, after which it would be discarded. It was perfect, except for the extreme hazard of getting the lander to *find* the return ship; a task requiring far more navigational accuracy than locating the proverbial needle in a haystack. But Houbolt's team argued that most manned activities in space would require very similar rendezvous procedures, so why not develop them immediately?

By November 1962, NASA had made their decision. Gradual occupation of earth orbit would be abandoned in favour of LOR. The ship for this project was designated 'Apollo'. A new and larger variant of Saturn would serve as a booster.

## ABANDONING STATIONS

Von Braun continued to argue against Houbolt's concepts for some while *after* he had recognized their engineering sense. He knew that Apollo would cancel out a sensible progression into space in favour of short-term achievements. His lifelong enthusiasm for space stations was founded on a simple instinct: no government would wish to leave a perfectly good and very expensive station uninhabited, and thus, a sustained space program would have to exist to keep it in service. The Apollo concept did not guarantee any such extension of space activities, but if the first lunar landing was to be achieved before Kennedy's deadline, then the

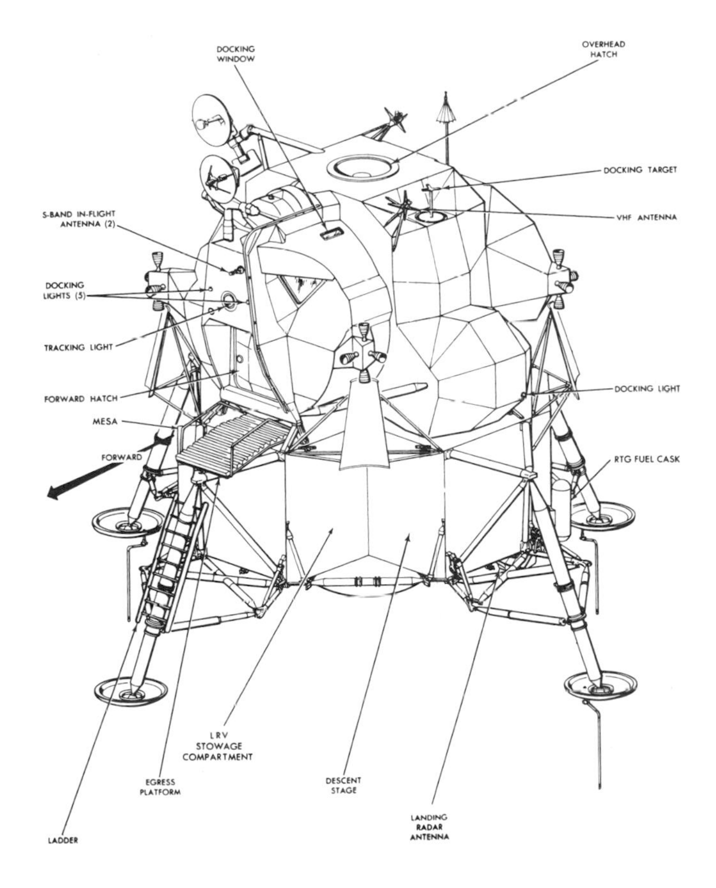


**FRAGILE BUG**
The Apollo Lunar Module was so fragile its legs could not support its weight on earth. The vehicle was secured during launch inside a Saturn by clamping the 'shoulders' of its legs against the interior of the protective shroud. The machine's solid appearance is deceptive. Most of its external structure was made from thin metal foil wrapped around a skeletal frame.

engineering argument for it was unassailable. Von Braun was too honest to stand in its way for long. Apollo–Saturn combinations eventually blasted up from the swampy flatlands of eastern Florida, then spent barely an hour in earth orbit before relighting their third stage engines and heading straight for the moon. The bug-like landing craft, the 'lunar modules', worked well, but their life-support capabilities were limited. This was the price that had to be paid in order to keep their weight down.

For all its flaws, Apollo remains a substantial achievement. Not least of its successes was the management of people as well as machinery. The political and commercial alliances that once conquered the moon have left their mark on modern NASA—often to the good, but sometimes not.

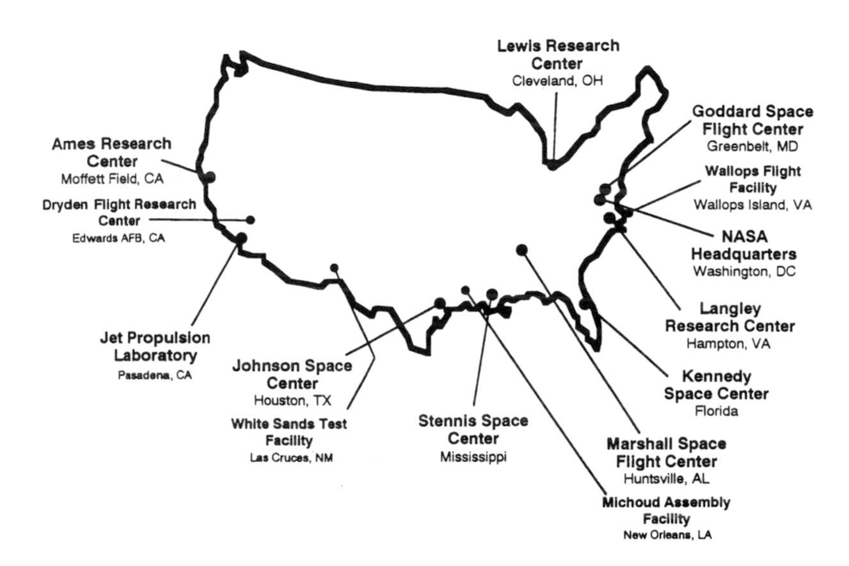


**SPACE TERRITORY**
An aerial view (*right*) shows the scale of one of NASA's large field centers, the Lyndon B. Johnson Space Center near Houston in Texas, from which all manned space missions are planned and controlled. The center employs up to 4,000 people at any one time, made up from permanent NASA staff working alongside contractors on short-term assignments. The Johnson Center benefits from the support of powerful Texan politicians, and often finds itself squabbling with other NASA centers over the distribution of funds and resources for new projects.

## THE POLITICS OF NASA

Regardless of their modest protestations to the contrary, most of NASA's chief administrators have wanted to leave their stamp on the agency, in the form of some grand new project or vehicle—with the possible exception of the man who helped establish America's space effort in the first place. Keith Glennan created NASA's basic structure from inception in 1958, although he never truly believed in the value of spaceflight. He was merely fulfilling Eisenhower's brief to achieve *something* in orbit without plunging America into reckless expense. But he created a swift-moving organization capable of planning for the future, even if he thought that the whole project should be limited. 'I never was a space cadet,' he admitted in later years.

Contrary to Eisenhower's expectations, rocket activities in Russia and America expanded swiftly. When President John F. Kennedy came to the White House in 1961, he and Vice-President Lyndon Johnson appointed James Webb as NASA's second administrator. Keith Glennan's cautious style was no longer deemed appropriate.

**NASA'S SCATTERED EMPIRE**

**Ames Research Center**
***Moffet, California* (1,700 staff)**
Performs general scientific aerospace research.

**Dryden Research Center**
***Edwards, California* (435 staff)**
Conducts test flight evaluations of prototype aircraft.

**Goddard Space Center**
***Greenbelt, Maryland* (3,820 staff)**
Responsible for development and flight control of unmanned planetary probe missions.

**Jet Propulsion Laboratory**
***Pasadena, California***
Staffed by contractors, JPL develops space probe systems.

**Kennedy Space Center**
***Merrit, Florida* (2,350 staff)**
Prepares and launches all manned and most unmanned rockets.

**Langley Research Center**
***Hampton, Virginia* (2,790 staff)**
Langley was established in the early years of this century to conduct aeronautical research, and it formed the basis for NASA's foundation in 1958.

**Lewis Research Center**
***Cleveland, Ohio* (2,460 staff)**
Investigates propulsion, control and power systems.

**Marshall Space Center**
***Huntsville, Alabama* (3,310 staff)**
Develops rocket engines, boosters and other large space structures.

**Michoud Assembly Facility**
***New Orleans, Louisiana***
Assembles shuttle fuel tanks.

**John C. Stennis Center**
***Stennis, Mississippi* (205 staff)**
Rocket engine test firing facility.

Jim Webb was a shrewd manipulator with many years of experience in the subtleties of Washington politics, and he assured NASA's long-term future on the back of President Kennedy's support. First, he deliberately overpriced the 1960s' Apollo moon landing project, telling Kennedy and Congress that it would cost $20 billion (a colossal sum by the standards of the 1960s) when his own advisors had costed it at $13 billion. Thus, by 1968, Webb could claim to be heading for the moon on time and on budget. In addition, he assured financial momentum from the emerging Apollo to prepare unmanned *Viking*, *Pioneer* and *Voyager* robot probes for missions to the planets, despite substantial cuts forced on space spending after 1967.

Many of NASA's physical facilities were constructed during Webb's tenure. He approved the huge Vertical Assembly Building (VAB) where rockets are stacked together at the launch center in Florida, and Houston's Mission Control Center—both of them significant icons of American culture to this day. It may seem odd that flights are controlled from Texas, while rockets blast off a thousand miles away in Florida, but this is Webb's deliberate legacy. In the early 1960s a powerful Texas politician was in charge of the congressional committee that presided over NASA's funding. Vice-President Johnson was another influential Texan. Webb turned Houston into the American space capital, and in return Apollo's budgets were eased through Congress.

Webb spread contracts far and wide in the 1960s so as to buy political support. His choice of the North American Aviation company (now known as Rockwell) to build the Apollo command module was controversial, and some historians have seen this contract as motivated more by political expediency than engineering validity; but Webb tried to counter such largesse with tight control of external contracts. Supervision of NASA was no less important to him. The various and widely scattered divisions within the agency had their own areas of responsibility, but everybody reported progress to a small group of senior managers at headquarters in Washington. Most important of all, there were only a half-dozen or so layers of rank within the agency, from the pad technicians in Florida right up to Webb's immediate colleagues. This flat ranking tree was remarkable for such a large federal organization.

**SATURN'S SHADOW**
A Saturn V booster on the launch pad in May 1973, prior to flying America's first space station, 'Skylab' into orbit. Saturn V was 110m (363ft) tall, and was powered at take-off by five F-1 engines, each delivering 680,000kg (1.5 million lb) of thrust. At the height of the Apollo–Saturn effort 500,000 people from around the American continent were involved somewhere in the program. Apollo's costs have often been criticized, but many workers would wish to see similar employment opportunities generated within the American economy today.

## TRIAL BY FIRE

Webb stepped down in October 1968, shortly before the first Apollo moon flights began, feeling that he'd done all he could after eight years. (He was 54 when he had first taken up his NASA post.) His bravest act as administrator had been to deflect onto himself, as far as possible, all the criticism for a fatal accident that occurred in January 1967. Three astronauts, Gus Grissom, Ed White and Roger Chaffee, died while their Apollo 1 spaceship was still sitting on the ground. The capsule was pressurized with pure oxygen, and a spark in an electrical conduit triggered an explosive flash fire. NASA had used pure oxygen in all their previous capsules without incident, and the engineers had become complacent about the risks. A number of flammable materials had been allowed into

**SMALL-SCALE SATURN**
Saturn 1B was 68m (223ft) tall and delivered a total thrust of 726,000kg (1.6 million lb) The 1B was the immediate ancestor of the huge Saturn V moon boosters. It was essentially a parallel cluster of smaller Redstone rockets bolted together within a common fuselage. The 1B was not powerful enough to reach the moon, but it was used to test Apollo capsules, and later, to fly crews up to Skylab. Its last flight was in July 1975, in support of the final flight of an Apollo. In this photo the 1B rests on a trestle next to the launch gantry usually reserved for a Saturn V.

**POWERING APOLLO**
James Webb was NASA's chief administrator from 1961 to 1968. His political skills helped the space agency to become the seventh largest federal organization in America, with a 5 per cent share of total national spending during 1966. Today that share is less than 1 per cent.

the ship. Most frightening of all, the capsule's hatch was poorly designed. Even under the calmest of conditions it took several minutes, rather than a few seconds, to get it open. Apollo 1's crew scrabbled at the hatch's levers, but to no avail. North American, the company that built the capsule, was criticized for their shoddy standards in electrical wiring and other assemblies. In turn, they accused NASA of being an impossible client, constantly revising the spacecraft specifications while it was under construction, and creating a climate in which mistakes could prevail. This was a deeply painful matter for Webb. He had intended NASA to serve as a proof of his credo, that huge technical and administrative problems *could* be solved by a government agency if only the management structures were right. But the fire had still happened, and for reasons that were less than forgivable.

## THEN AND NOW

James Webb's reputation today rests on his positive achievements, his old-fashioned faith in the potentials of good government, and on his essential nobility of character. But as any number of NASA chiefs since his day will attest, he had a good deal of luck riding with him. Firstly, his mandate was clear: to put America on the moon before 1970. Secondly, he never had to worry about money. NASA was granted 5.3 per cent of the entire federal budget at a time of great economic optimism, before Vietnam, social upheaval and global recession during the 1970s changed the American outlook. The current space chiefs live in a very different world. They can't be as certain as Webb about what their country is asking them to achieve, and they spend their time worrying about little else except money. NASA's share of spending for 1994 amounted to 0.9 per cent; less than $15 billion. (Webb's average budget of $5 billion a year in the mid-1960s would equate to $30 billion in today's dollars.)

And it gets worse. Budget pressures may force the agency to reduce spending by $4 billion a year before the turn of the century. Meanwhile, the national economy as a whole struggles with a deficit counted in trillions of dollars. In such a tense financial environment the space agency's annual budget is regarded as a soft target.

**LAST OF THE FIRST**
The Russian Salyut 7 space station, the longest-lasting of the Salyut series, here seen in orbit from a departing Soyuz ferry after a crew transfer in September 1985. Another Soyuz remains attached to the station. Launched on April 19, 1982, Salyut 7 remained operational until June 1986. The station was 15m (49.2ft) in length, and weighed 18,500kg (41,674lb). It carried front and rear docking ports and was equipped with twin rocket thrusters delivering 300kg (661lb) of reboost power.

Perhaps the greatest change since Webb's day is that big federal agencies such as NASA are no longer trusted by the general population, nor even by their own political masters. In recent times, a pervasive anti-government mood has taken root. During the 1980s NASA's management structures were held in some contempt. Critics said the agency had grown middle-aged and fat. Sadly, this is what tends to happen to most large organizations over time. The youthful fire of ambition tends to be replaced by the more cautious desire to protect hard-won territory.

The best thing that ever happened to NASA in their early years, and something which they badly needed again, was the presence of a powerful competitor to sustain their momentum. American astronauts may have dipped into the cosmos as fabulously well-equipped tourists, but Russian cosmonauts in simpler, cheaper ships made space their second home.

## RED STAR IN ORBIT

Throughout the 1960s NASA's competitors in the Soviet Union were anxious to develop a Saturn-class booster of their own, though they didn't make this clear to the outside world until 25 years later. On February 21, 1969, an American spy satellite took snapshots of a new Soviet launch complex with a large and hitherto unidentified rocket standing upright alongside the tall gantries. American intelligence experts feared that Russia was about to make an unmanned test flight of a moon booster. On the spy probe's next pass, the booster was gone—and so was the launch gantry. In 1995, Russian officials finally admitted that their 'N1' rocket had exploded shortly after ignition, thus ending their hopes of beating Apollo to the moon.

The Soviets fell back on using smaller but more reliable rockets to exploit the possibilities of earth orbit. To them goes the honour of having built and launched the world's first space station. On April 19, 1971, 'Salyut' ('Salute') was launched atop a Proton booster. It was a cramped, stuffy cylindrical compartment, barely 3m (9ft) in diameter at its widest point and 15m (49ft) long. The interior space compared unfavourably with that of a Greyhound bus filled with luggage, but Salyut provided more living room than any space travellers had experienced before.

Three days later a 'Soyuz' crew ferry made a rendezvous with it, although the trio of cosmonauts inside the cramped capsule did not transfer to the station. Their docking system could not achieve a proper seal. On June 6, cosmonauts Georgi Dobrovolsky, Vladislav Volkov and Viktor Patsayev lifted off in another Soyuz to become the first occupants of a home in space, living aboard Salyut for 23 days. As it turned out, those days would be precious indeed. The Soyuz ferry was still a temperamental vehicle. After its first test flight on April 24, 1967 the single test pilot, Vladimir Komarov, was killed when his reentry capsule's parachute became entangled. Coming barely three months after NASA had lost Chaffee, Grissom and White in the Apollo 1 fire, Komarov's crash reminded everybody just how dangerous rocket travel really was. Politicians continued to play up the competitive element of the Space Race, but fliers under both regimes now felt a sense of kinship.

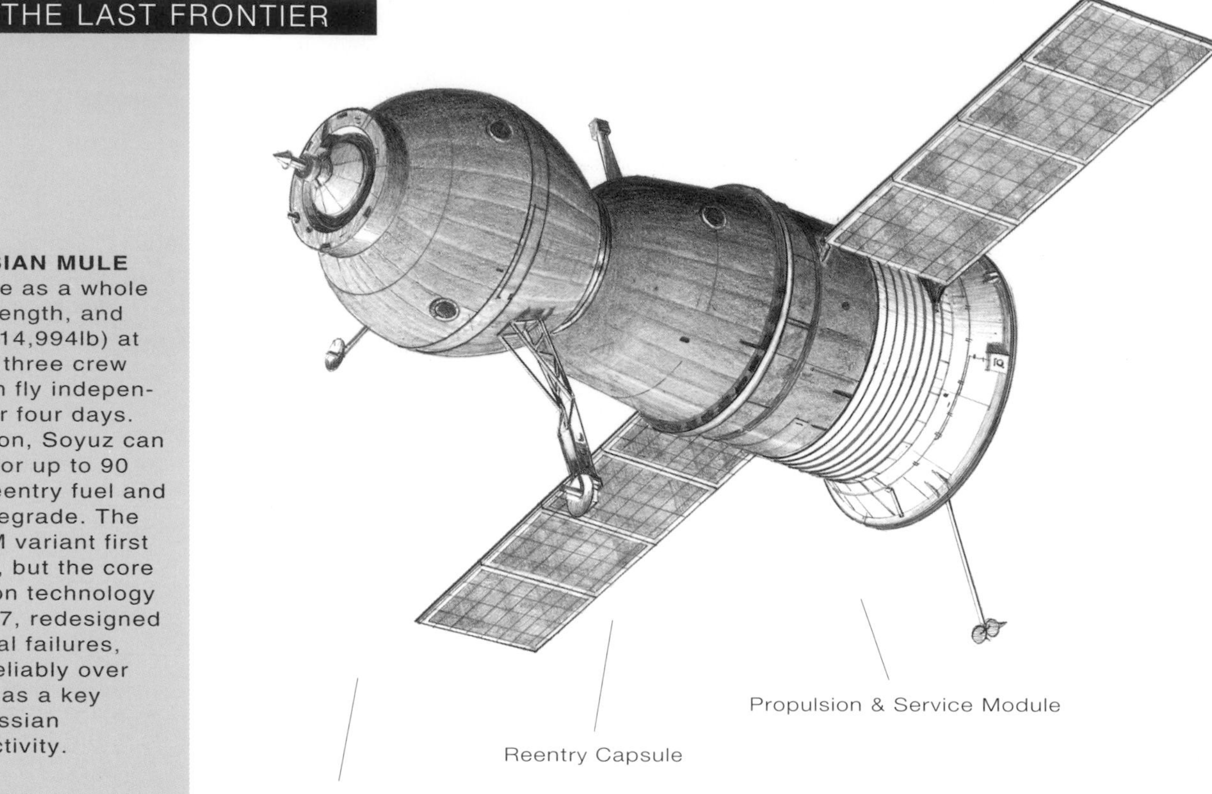


**FAITHFUL RUSSIAN MULE**
The Soyuz vehicle as a whole is 7.9m (26ft) in length, and weighs 6,800kg (14,994lb) at launch. It carries three crew members and can fly independently for three or four days. Docked to a station, Soyuz can remain in space for up to 90 days before its reentry fuel and power supplies degrade. The current Soyuz-TM variant first flew in May 1986, but the core design is based on technology developed in 1967, redesigned in 1971 after initial failures, and since used reliably over the last 30 years as a key element of all Russian manned space activity.

Since Komarov's accident, further flights of the Soyuz seemed to have proved the system, but on June 29, 1971, when Salyut 1's crew unlatched from the station to begin their reentry, all the air leaked out of the cabin through a faulty valve. By the time their return capsule had parachuted back down to earth, Dobrovolsky, Volkov and Patsayev were dead.

The Soyuz system had to be redesigned, and it was not possible to launch another crew to Salyut 1. Six months after its launch, the station's orbit had weakened. It fell into the atmosphere, burning up in a remote-controlled suicide assisted by ground controllers. The Russians improved their Soyuz capsules (they are still in use today) and launched six more Salyut stations between 1972 and 1982, keeping them in stable orbit for ever longer periods and regularly sending up fresh crews. The biggest item of equipment on board several Salyuts was a high-resolution camera, through which the cosmonauts dutifully peered down at NATO exercises and other Western military activities. Official Soviet press reports mentioned 'earth observations in the interests of the national economy'.

Genuine science was never entirely forgotten, however. The Russians gained substantial experience in astronautical medicine, biology, astronomy, metallurgy, and in the myriad activities that make up an active life in space. From 1975 they learned how to resupply their Salyuts with unmanned robot ships, 'Progress' ferries loaded with fresh oxygen, food and water. Progress was adapted from Soyuz hardware, with the crew cabins stripped out to make room for cargo. It was an efficient piece of Russian design, building on what they already had.

During February 1985 the final Salyut station had been temporarily unoccupied for a few weeks, and another crew was scheduled to go up and join it. But the station's solar panels had lost their orientation to the sun. With the navigation systems starved of electrical power, Salyut 7 gradually drifted out of control. Meanwhile, the air conditioning systems failed and the interior cabin space filled with toxic fumes. Down on the ground, cosmonauts began training for a rescue mission. On June 16, Vladimir Dzhanibekov and Viktor Savinykh warily piloted their Soyuz capsule around the flailing station like boxers circling just beyond the reach of an angry opponent. Instead of abandoning the mission, they worked out a way of docking and then made repairs.

This last Salyut hosted guest astronauts from other Communist countries, and even from France and India. Soviet envoys happily offered flight time on board their station to any Americans who might want it. Washington officials politely declined through gritted teeth, and prayed they could get their own space station developed some day soon.

A *second* space station. The first one was called 'Skylab'.

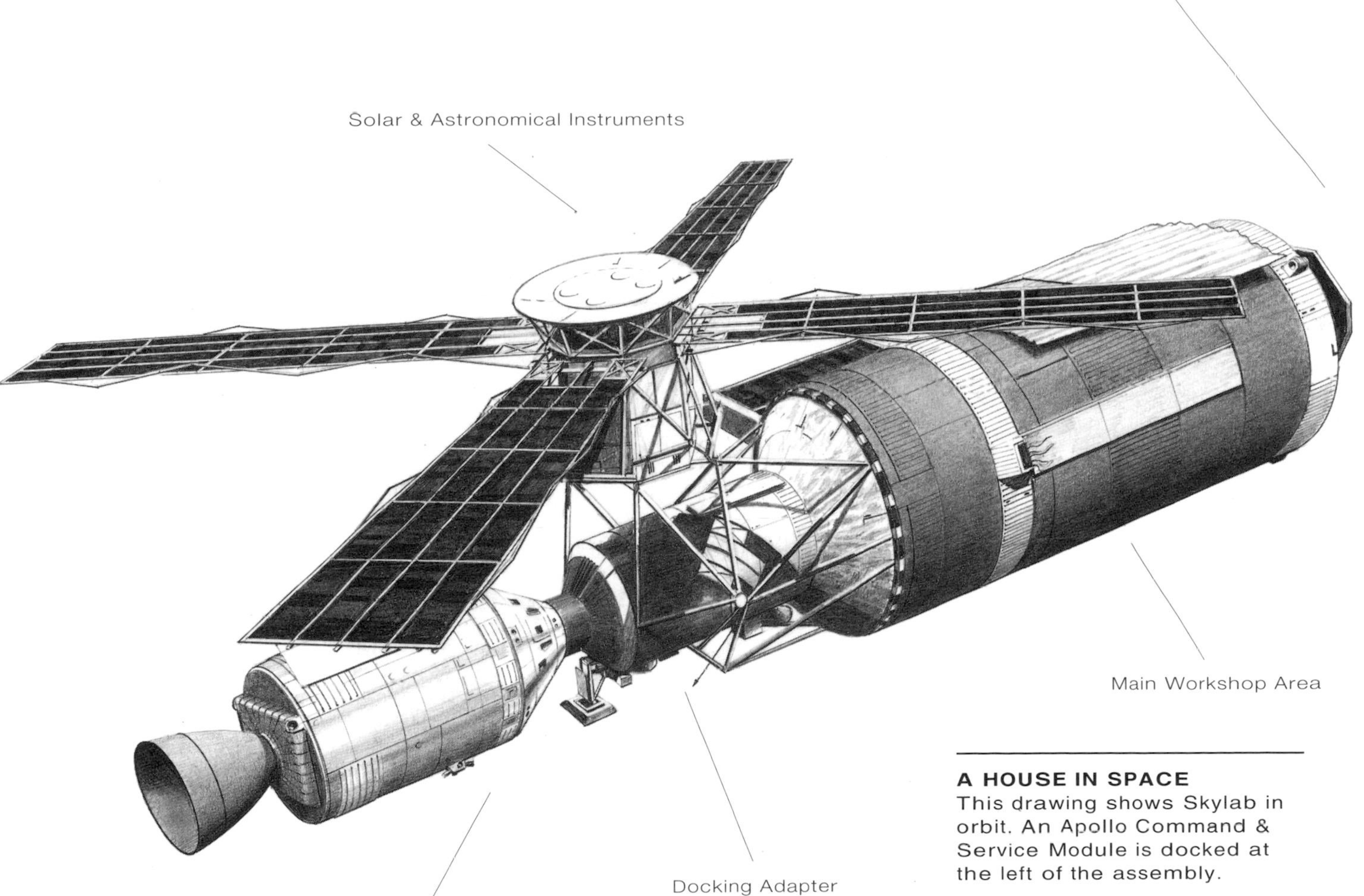


**A HOUSE IN SPACE**
This drawing shows Skylab in orbit. An Apollo Command & Service Module is docked at the left of the assembly.

**BROKEN, BUT FLYING**
Skylab in orbit during February 1974, seen from a departing Apollo capsule, the last ship to visit. A crumpled sheet of gold foil has been erected to supplement thermal cladding damaged by vibration and atmospheric drag during Skylab's unmanned launch on May 14, 1973. The left solar sail was completely torn away. With a docked Apollo, Skylab measured 36m (118ft) from end to end, and weighed 89,630kg (197,600lb).

**FIRST CREW**
Charles Conrad *Commander*
Joseph Kerwin
Paul Weitz
**Crew launch:** May 25 1973
**Crew return:** June 22 1973
**Mission duration: 28 days**

**SECOND CREW**
Alan Bean *Commander*
Owen Garriott
Jack Lousma
**Crew launch:** July 28 1973
**Crew return:** Sep 25 1973
**Mission duration: 59 days**

**THIRD CREW**
Gerald Carr *Commander*
Edward Gibson
William Pogue
**Crew launch:** Nov 16 1973
**Crew return:** Feb 8 1974
**Mission duration: 84 days**

## SKYLAB

By 1970, NASA's space activities were losing support from the general public, in the wake of more pressing national concerns. Apollo 20 was supposed to have been the final lunar trip. It was cancelled, along with flights 18 and 19. Some space administrators were secretly relieved. In April 1970, Apollo 13 was nearly lost in space after an oxygen tank exploded when the ship was already halfway to the moon. NASA's recovery of the crew showed remarkable powers of management and technical ingenuity, but the agency chiefs were reasonably happy to shut down the moon program before anything worse could happen.

Two Saturn boosters intended for cancelled lunar flights now form probably the largest and most expensive museum exhibits in the world, dismembered, lying on their sides like gigantic beached whales at space centers in Houston and Huntsville. (The version on display in Florida was a test vehicle, never intended to fly.) But a third spare Saturn was kept on flight-ready status. NASA had always hoped to use their Saturns for something called 'Apollo Applications', a program intended to exploit the launch hardware nearer earth once the lunar missions were completed. A space station of some kind remained one of the agency's favourite ambitions, even though its purpose was becoming increasingly vague. Obviously it could not be a staging post to the moon, since that problem had already been solved.

**OUTSIDE THE LAB**
August 1973: astronaut Owen Garriot replaces a solar particle measuring device on Skylab's telescope assembly.

**A RARE OFF-DUTY MOMENT**
January 1974: Gerry Carr (*left*) spins Bill Pogue on his finger during the third and last Skylab mission.

In the mid-1960s NASA still imagined that they might be allowed to fly a Mars mission sometime before the end of the century, perhaps using components lifted into space on a series of Saturns. Meantime, it seemed sensible to study how human beings adapted to long periods in space. These plans were speculative at best, since the agency had not so far obtained firm backing to use Apollo hardware for anything more than the actual lunar missions. Most of their longer-term goals were reined back, but after 1969 there remained sufficient funding for at least one idea on the Apollo Applications wish-list: a project called Skylab.

## NOT FIRST, BUT BEST

The big factor in Skylab's favour was its economy. For all intents and purposes, it was already half-built even before the program was formally approved. The third stage of the last available flightworthy Saturn (for various complicated reasons this stage was known as a Saturn S-IVB, 'S-Four-B') could be hauled into orbit atop the main booster, but with its cavernous tanks left empty of fuel. Lightweight flooring and dividers could be used to convert the airtight tanks into cabins. The launch weight not taken up by a moon-bound capsule and lunar lander could be allocated for different types of equipment suited for a space station. Three half-size variants of Saturn left over from early test programs could then be used for launching Skylab's crews in spare Apollo capsules. NASA engineers had recognized these possibilities since 1965, after a senior Apollo manager, George Mueller, had sketched a rough outline on a paper napkin during lunch one day. In 1969 President Nixon's intelligence advisors suspected that a Soviet space station was under development, and a response had to be formulated. The McDonnell Douglas company was awarded the contract to convert a Saturn S-IVB third stage into Skylab. Their selection as contractor came down to common sense. They had fabricated the S-IVB stages in the first place. The conversion was completed in 36 months for a budget of $2.5 billion.

Of course, the first Soviet Salyut got into orbit in April 1971, two years ahead of Skylab, but NASA's people comforted themselves with the knowledge that their station would be the largest habitable piece of machinery ever put into space, incorporating gigantic solar panels, a mass of valuable equipment, a large astronomical and solar telescope array, dozens of scientific and medical experiments loaded into the many interior storage lockers, and a cylindrical working compartment as roomy as a two-bedroomed house.

The last of Wernher von Braun's Saturn Vs ever to fly thundered into orbit on May 14, 1973, carrying Skylab. The launch appeared to be flawless, and the 75-ton station assumed a perfect orbit, circling earth at a height of 435km (270 miles) just as planned. Unfortunately, when the time came for Skylab to deploy its various systems under remote control, a number of serious failures were revealed. Micrometeoroid shielding on the skin of the converted S-IVB had been torn out of place during flight, and one of the two main solar panels was jammed shut in its pre-launch position. The second solar panel was completely torn away like a bird's wing ripped off at the shoulder.

In those days, NASA had a lucky knack for turning problems into triumphs. Just as they had gained public favour for bringing Apollo 13 back from the brink of tragedy in 1970, so Skylab presented another newsworthy drama. The first crew would have to make a dangerous spacewalk, effect repairs and stabilize the drifting station before it became uninhabitable. Failure would be a costly embarrassment, but success would enable NASA to 'prove' the value of putting astronauts into space.

Launch of the first crew was postponed. Astronauts Joe Kerwin and Paul Weitz, along with their commander, Pete Conrad (a Gemini and Apollo veteran) embarked on a 10-day crash training program, testing out various *ad hoc* fixes for Skylab in a watertank spacewalk simulator at Houston, rapidly reinventing a mission for which they had already spent many months in training. On May 25, 1973, Conrad and his crew docked with the crippled lab and conducted an urgent series of spacewalks to free

**A CLUTTERED CAN**
Skylab's main work area was a roomy cylinder 14.7m (48ft) long and 6.6m (21ft) in diameter. The partitions were made from a grid with triangular gaps, so that the crew could secure themselves by pushing toggles on the soles of their boots into the holes in the flooring.

the solar panel, and to erect a simple sun shield in an effort to lower the lab's interior temperature. Their repairs were successful, and they went on to spend a month in orbit, bringing the station's systems on-line and conducting experiments. Skylab was well equipped with scientific payloads, including an array of instruments for observing the sun in unprecedented detail. In July another crew visited Skylab for 59 days. The last crew, Gerry Carr, Bill Pogue and Ed Gibson, splashed down on February 8, 1974, after an 84-day stay aboard the station.

By 1979 Skylab had been drifting unoccupied for five years, and its orbit was degrading fast. On July 11 it plunged into the atmosphere entirely out of control. Space officials crossed their fingers and hoped that the debris would fall harmlessly into the sea. In the event, burning fragments scattered across hundreds of square miles of Australian outback. One cow was killed when a lump of the station fell on its head.

After 1974, with the Skylab program completed, some of NASA's older astronauts feared they would never have a chance to fly again. A new generation of space vehicle was under development, but it would not fly for another seven years. Meanwhile, only one further Apollo mission remained on the schedules. It wasn't a trip to the moon, or even another session aboard the abandoned Skylab. It was just a simple hop into orbit for a few days' fraternization with the enemy.

## THE APOLLO–SOYUZ TEST PROJECT

President Nixon was no great supporter of NASA, but he encouraged the first joint space mission between America and Russia, thus bringing to an end a decade of breakneck competition. The Apollo–Soyuz Test Project (ASTP) stemmed from tentative meetings between NASA and Russian space chiefs during October 1970. In 1972, Nixon and President Alexei Kosygin of Russia signed a formal space pact that allowed Russian and American space engineers to establish a Joint Working Group. This was a diplomatic breakthrough, and it seemed to augur well for future East–West relations.

The Soyuz ships were in standard use by now, regularly docking with Salyut stations. Meanwhile, America's fleet of spaceships was nearly expended. After making allowances for putting crews aboard Skylab, NASA would be left with one usable Apollo command module and a final flightworthy Saturn—the half-size version. The Working Group decided to dock this Apollo with a Soyuz. In terms of scientific knowledge, or advancing the exploration of space, such a mission would be almost entirely without merit. Politically, it would reap some rewards.

ASTP's planners realized that persuading engineers from the rival superpowers to think along the same lines would be an achievement in itself. Apollo and Soyuz used different orbits, different docking systems, different boosters, different cabin atmospheric pressures.

The solution was a small, jointly designed docking adapter that attached to the nose of the Apollo in the place where a lunar module might normally have been. The 'open' end of the adapter carried a system suitable for mating with the nose of a Soyuz. The adapter also contained

its own atmospheric regulation system to act as a buffer between the two ships' atmospheres and to equalize the pressures once docking had been completed. The technical problems proved less complex than the diplomatic and security implications of sharing the docking technology.

On July 17, 1975, the docking took place. The two crews shook hands, clambered into each other's ships, and shared food and drink. Genuinely moved, Russians on the streets of Moscow gave great bear hugs to foreign visitors. It seemed a breakthrough for peace, yet it would take 20 years for Russian and American manned space technology to get this close again. The new political détente proved fragile.

Apollo was commanded by Tom Stafford, a veteran astronaut. His two crewmates were 'Deke' Slayton and Vance Brand. The Soyuz commander was another veteran flier, Alexei Leonov, who was the first man to walk in space in 1965. Valeri Kubasov was his co-pilot.

ASTP was flawless, until the moment when NASA's last Apollo splashed down into the Pacific to await recovery by a Navy aircraft carrier. Excess thruster fuel leaked into the spacecraft's cabin and poisonous fumes seared the astronauts' lungs. Vance Brand lost consciousness. Fortunately, no permanent damage was done to the crew's health.

With ASTP completed, all the Apollo hardware was used up. NASA astronauts were grounded for the next six years, waiting for a new space vehicle to fly. Meanwhile, Soyuz ships continued to fly on a regular basis as the Russian space station program continued its steady advance.

**APOLLO MEETS SOYUZ**
An artist's impression of the docking between the last of America's Apollo capsules (*left*) with a Russian Soyuz on July 17, 1975. The Docking Adapter is attached to the Apollo's nose.

## SPACE STATION MIR

On February 19, 1986 the Soviets launched the first component of their new-generation 'Mir' ('Peace') station. Today, it is still in orbit along with a myriad of additional compartments that have been added to it over the years. Mir now regularly hosts astronauts from Western countries—including America. In the wake of the collapse of the Soviet empire after 1990, a very different mood has prevailed in space.

On March 13, 1986, Mir's first crew, Leonid Kzim and Vladimir Solovyev, blasted off for a rendezvous and then clocked up an impressive roster of activities. After two months aboard Mir they took their Soyuz for a 50-day visit to the old Salyut 7, which was still in orbit. They made two space walks, investigating the exterior condition of Salyut and practising assembly work in space. Then they flew back to Mir, reoccupying the new station for three weeks before finally heading for home after a mission totalling 125 days. By any measure, the Soviets had consolidated their superiority in earth–orbital operations using relatively inexpensive standardized technology to achieve feats that NASA had little immediate prospect of matching.

In the course of such accomplishments, Russian cosmonauts have laid claim to all the duration records: the longest-functioning manned space vehicles; the greatest distances travelled in orbit; the greatest number of crew changes; the greatest total accumulated experience in space; and time and again, the longest periods spent aloft by individual crew members. This last record belongs to Dr Valeri Poliakov, who spent 438 days—well over a year—in orbit aboard Mir. During the final stages of his mission he was joined by Yelena Kondakova, who spent six months on the station and became the longest-serving woman space traveller. Poliakov came home in March 1995. His main problem was not so much the long mission itself, but readjusting afterwards to life back on earth.

American observers are often bemused by Russia's apparent fascination for endurance records, with specific scientific work often taking second place to simply clocking up time. One of NASA's officials recently likened the cosmonauts' dull routine on Mir as 'just a process of logging up endless days, boring holes in the sky. Their job is basically to withstand being up there.'

Even as the old Soviet order collapses; even as their space activities are cut to the bone; even with no immediate prospect of achieving their long-term goal, it remains politically acceptable for Russian space workers to plan for the impossible. They are attempting to gain experience for a flight to Mars. Their cosmonauts are trying to prove that they could survive mentally and physically, in cramped,

**FIRST YEAR IN SERVICE**
The core module of the Russian space station Mir during 1986, with a docked Soyuz ferry.

weightless cabins through such an immense journey. For our erstwhile Cold War enemies, it seems a curiously hopeful and touching ambition to try and conquer the far planets when they can't even feed themselves properly on the ground. A NASA astronaut lamented recently, 'For the Russians, space is still an important icon of national culture, despite all their economic problems. For us, it's just another shop in the mall.'

Meanwhile, what had America's space agency been doing since the end of Skylab and the ASTP mission of 1975? Trying to secure their future by designing a new kind of space vehicle.

**A DECADE LATER**
A distant view of Mir in 1996, showing clear evidence of expansion since 1986.

## THE 'WORKHORSE'

James Webb's successor as chief of NASA was Tom Paine, appointed in 1968. His principal legacy was the Skylab project, but he also set America on the road to developing the 'Space Shuttle'.

Paine found the prevailing political climate very different from Webb's day, and in some ways he wasn't ready to accept that NASA had to change. He thought that the lunar landing missions had set a precedent for further large-scale space projects, but he was wrong. Apollo had been the exception, not the rule. It was Paine who trimmed the tail end of the moon program, trying to save money and appease the new president of 1969, Richard Nixon.

Paine acknowledged that American taxpayers would no longer tolerate writing off expensive Saturn rockets after each flight, but he believed that NASA had a solution to this wastage. Their proposed shuttle was designed to be as reusable as an airplane. Paine hoped that Nixon's administration would allow the savings from cancelled Apollo flights to be allocated for development of the shuttle, just like the sleek spaceplanes in one of his favourite paintings, Chesley Bonestell's opening illustration for the first of the *Collier's* space articles in 1952. Unfortunately, Paine was no match for Nixon's advisors, and he resigned after just two years. (In

## FLAGSHIP SPACE TECHNOLOGIES: OLD, UNRELIABLE AND EXPENSIVE

Russia's Mir space station (above) has been in orbit for more than a decade. The addition of new modules has added to the station's working life and greatly expanded the interior space, but the core elements launched in 1986 are no longer reliable. Russia can no longer support Mir's costs without foreign help.

Meanwhile, America's space shuttles are also beginning to show their age. Pictured here is the *Atlantis* lifting off from the Kennedy Space Center in Florida on June 27, 1995, bound for a rendezvous with Mir. The oldest shuttle, *Columbia*, has been flying since 1981. The main rocket engines rely on technology developed in the 1970s. NASA will remain dependent on shuttles until a replacement system can be developed and funded, perhaps sometime in the next 8–10 years. The ideal launch vehicle will be a single-stage system, easy to maintain and with no throwaway components.

USA
NASA
Atlantis

1986 he explained to a journalist, 'I finally left because I didn't think I could deliver the kind of relationship with the President which the head of NASA really ought to deliver.')

NASA's next administrator, Jim Fletcher, had hoped for $10–15 billion to build the shuttle. Under political pressure he revised his estimate to $8 billion. This was still too expensive for the White House, and they asked for another and yet lower budget to be submitted for their approval—until Caspar Weinberger, a senior Republican politician, intervened to persuade President Nixon that NASA's requests couldn't be shelved indefinitely. Some kind of commitment had to be made. On August 12, 1971, Weinberger sent Nixon a memo, a single short document that probably saved the space shuttle program from a lingering death of a thousand cuts. Weinberger argued there was a danger of giving the impression that 'our best years are behind us, and that we are turning inward'. He further stated, 'NASA's proposals have some merit ... The real reason for sharp reductions in NASA's budget is that we cut it because it is cuttable, and not necessarily because it is doing a bad job.'

## GRUDGING APPROVAL

Nixon was in a difficult position. He could not count on public support for a big NASA program, but neither did he want to kill the space effort altogether, because that would have been damaging to America's international prestige, which mattered a lot to him since he prided himself on his agenda for foreign diplomacy. So he scrawled a brief note in the margins of Weinberger's memo, 'I agree with Caspar'. However, by January 5, 1972, when Nixon formally announced the shuttle program, the permitted budget had shrunk to $5.5 billion, a 30 per cent shortfall in NASA's 'lowest option' funding even before the first pieces of hardware were built. NASA was forced into devising a fudged system with a throwaway fuel tank and twin solid rocket boosters, instead of their ideal design, an elegant 'piggyback' vehicle with a winged booster that could land and be reused just like the main spacecraft itself. A system that could have been very cheap to run, if only the initial investment had allowed for a proper design, was supplanted by something that was cheap to build. NASA ended up with an imperfect piece of technology.

Everbody at NASA had imagined the shuttle would be just one element of that old dream, an orbiting infrastructure. What was the *point* of a shuttle if it had nowhere to go? Many detailed space station designs had been formulated, but no serious budget requests were put forward. The agency gambled on 'proving' their shuttle concept first, and then winning funds for a space station in future years.



Confronted by harsh financial realities, Jim Fletcher introduced what seemed at the time a bold and practical initiative, a plan to make the shuttle available as a 'workhorse' for private industry as well as for space scientists. NASA would attempt to offset a proportion of the costs by charging fees for launching other people's payloads. Fletcher reinforced the fragmented political support for a continued manned space program by promising cheaper access to space for private communications satellites, university experiments and military payloads. This was certainly an appealing prospect for Nixon, and for subsequent presidents. It seemed a way of making NASA a genuine national resource instead of just an expensive club for space cadets. Critics argued, with some justification, that there was no reason to carry astronauts and life-support equipment in the *same* vehicle as an unmanned satellite, which could always be launched more cheaply in a small expendable rocket. But NASA suggested that the shuttle would represent a perfect compromise between manned and unmanned operations, a 'do-it-all' spaceship.

## FLETCHER'S FIGURES

James Fletcher relied on sophisticated cost-benefit analyses to back up his arguments. He hoped to eliminate all NASA's ageing throwaway rockets. But the shuttle hadn't even been built yet, and nobody could properly claim any previous experience operating such a machine because nothing like it had ever flown before. Inevitably, all the analysts' figures about its anticipated cheap operating costs and reliability were speculative at best. NASA approached private companies, universities and research institutes, trying to interest them in buying future payload space aboard the shuttle. At one point, a fee of $220 per kilo ($100 per pound) was predicted, with a shuttle launch rate of 50 or 60 flights per year, divided between a fleet of 5 orbiters. These figures would prove to be unrealistic. NASA would achieve less than 1 flight a month, and payload prices would eventually work out 50 times more expensive than Fletcher's analysts had forecast.

Throughout the 1970s there were several changes of administration in the White House. The Watergate scandal ended Nixon's occupancy in 1974; Gerald Ford's caretaker presidency gave way to Democrat Jimmy Carter, who was subsequently crippled by the Iranian hostage crisis. All these political dramas overshadowed NASA's fortunes.

On April 12, 1981, morale within NASA was revived when the first operational space shuttle, *Columbia*, blasted into orbit, piloted by veteran astronaut John Young and newcomer Bob Crippen. American astronauts were in orbit again after a furlough of six years. The shuttle seemed to work well. Public approval for space flight reached a new high after a decade of indifference. Best of all for NASA, the latest man in the White House—Ronald Reagan—began to show a keen interest in their new ship. Within his first term in office, he would give his blessing to the next big prize for the space cadets: a giant orbiting station.

But first, NASA's chiefs had some hard selling to do.

# FIGHTING FOR FREEDOM

*In 1984, President Reagan announced his support for a new and permanent space station, intended for completion in the early 1990s. Across the next eight years, NASA spent $8 billion on a complex series of design reviews, but no elements of the station were fully completed, let alone put into orbit. In 1988 the project was given a name: 'Freedom'.*

**ONE OF MANY VERSIONS**
In this 1991 painting, the Freedom space station consists of pressurized modules attached to a central truss beam, as seen against a bright backdrop of cloudy sky. This design was not the first, nor would it be the last.

# FIGHTING FOR FREEDOM

Ronald Reagan became president in 1981, sweeping aside Jimmy Carter's beleaguered administration by a comfortable margin at the polls. Reagan's approach to space became the most adventurous since that of President Kennedy in the 1960s. As it turned out, he was thinking very big indeed. On March 23, 1983, he made a startling public announcement: 'Until now, we have based our nuclear deterrence on the threat of retaliation. What if we could live secure in the knowledge that we could intercept enemy missiles before they reached our soil?' Reagan was promoting the Strategic Defense Initiative (SDI). A network of 'killer' satellites, laser platforms and other exotic weapons would girdle the globe, protecting the 'Free World' against what Reagan once called 'The Soviet Evil Empire'.

Edward Teller, inventor of the hydrogen bomb (and the real-life inspiration for the movie character Dr Strangelove) had sold Reagan a pipe dream. Most scientists were doubtful that SDI, 'Star Wars', could be made to work, but military spending under Reagan escalated to levels not seen since the height of the Vietnam War, and Teller's supporters had little difficulty initiating big new projects. A number of high-technology companies made ambitious claims for SDI in order to receive huge research grants.

It might be argued that SDI 'worked'. The Soviet Union could not hope to match the research spending, and this had a significant impact on the easing of the Cold War by the end of the 1980s. The truth remains that SDI was an expensive technological lie. Some companies preloaded their target missiles with explosives so that they would appear to blow up when 'hit' by laser beams, interceptor weapons and so forth that were, in fact, very wide of their mark. Reagan may not have been overly concerned with the technical details, but he made effective use of SDI's apparent threat in his subsequent peace negotiations with Russia.

The Pentagon's SDI plans were real enough, however. Vandenberg Air Force Base in California was adapted to handle shuttle launches and landings for defence-related missions, although the military fell somewhat short of paying for a shuttle of their own. Instead, they persuaded NASA to allow them access to flights at short notice in return for dubious political support. Civilian space staffers were unhappy with this arrangement, particularly as NASA's original 1958 mandate as a purely civilian agency had encouraged openness and non-aggression in all matters. But a significant proportion of shuttle flights was now committed to carrying



**THE LOGICAL NEXT STEP?** By the mid-1980s NASA's design for their planned space station was this 'dual keel' configuration. At its widest point the station would have spanned close to 150m (500ft). Astronomical experiments and non-pressurized payloads are attached to the truss elements, and a shuttle is docked. The array of hexagonal panels at the top of the picture consists of steerable solar furnace mirrors which focus the sun's rays and create electrical power more efficiently than the main solar panels. The blank rectangular panels are radiators.

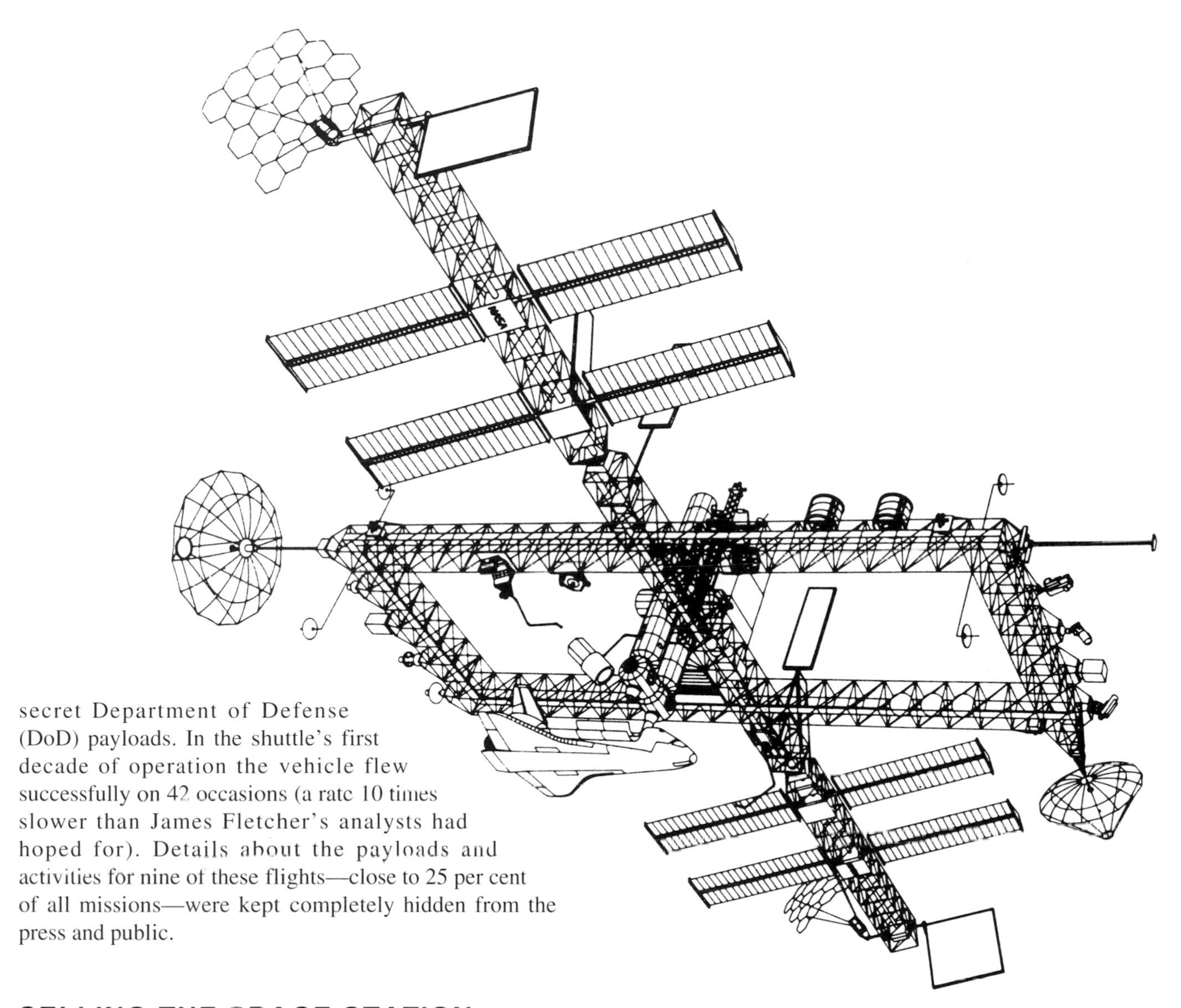

secret Department of Defense (DoD) payloads. In the shuttle's first decade of operation the vehicle flew successfully on 42 occasions (a rate 10 times slower than James Fletcher's analysts had hoped for). Details about the payloads and activities for nine of these flights—close to 25 per cent of all missions—were kept completely hidden from the press and public.

## SELLING THE SPACE STATION

In June 1981, the new Reagan administration put James Beggs in charge of NASA. During the 1970s Beggs had served under President Nixon as secretary of transportation, and his career had also encompassed executive positions with some of America's most prominent corporations. Beggs's deputy was also chosen by the White House: Hans Mark.

Mark had been a senior Air Force staffer for President Carter in the 1970s, and he was keen to run NASA himself. But Beggs was a true Republican supporter, which gave him the edge with the Reagan White House. Mark's strength was his detailed knowledge of the defence establishment. He had a boyish enthusiasm for the space shuttle, and had done much of the work involved in placing the DoD payloads into NASA's launch schedules. He saw himself as a peacemaker between the civilian space community and the military. Not everybody at NASA agreed with him on that point but, as an erstwhile student of the H-bomb expert

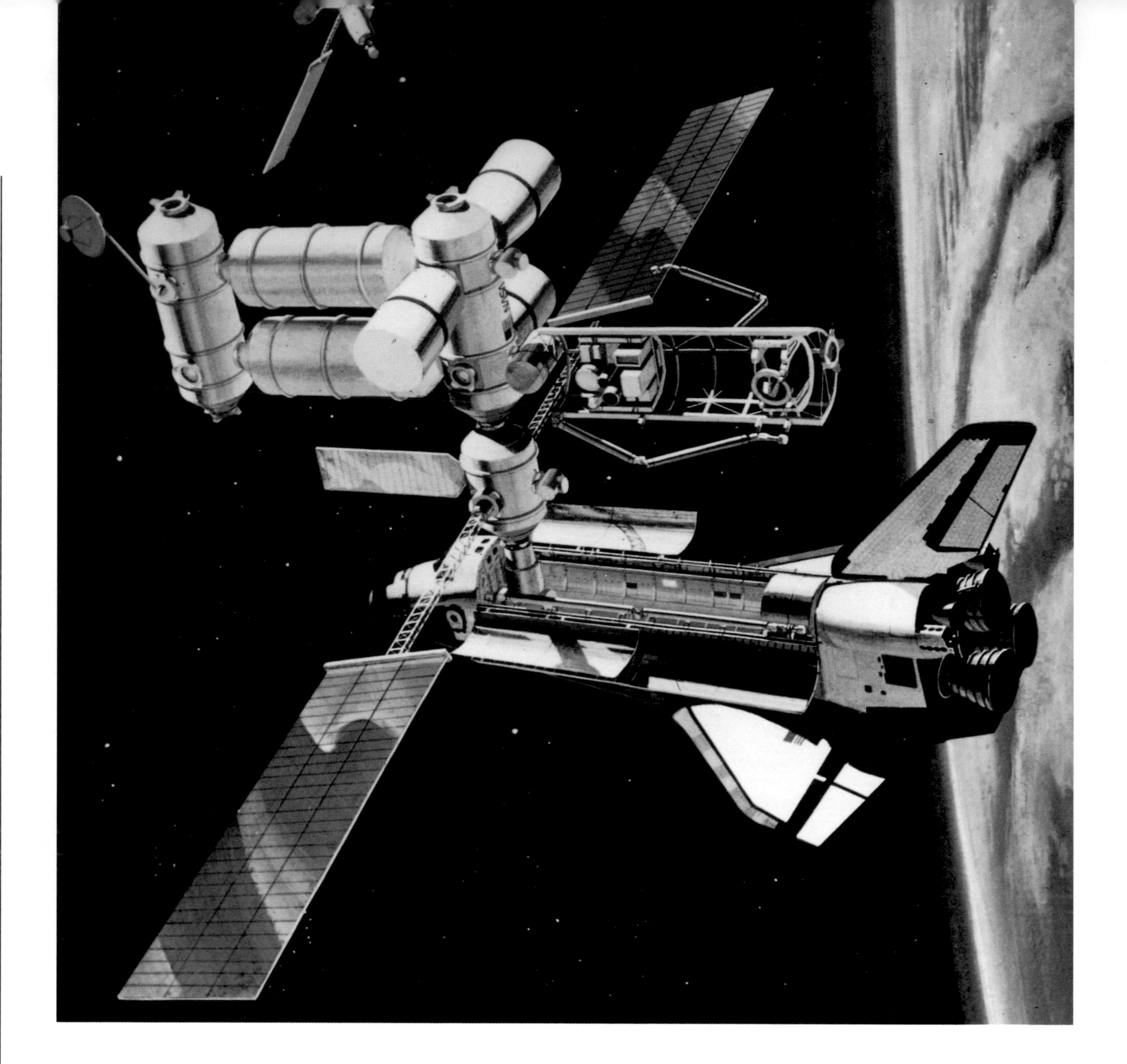

Edward Teller, Marks thought he would be able to sway the SDI supporters at the White House to be friendlier towards space. While the shuttle system began to prove itself after the first flight of *Columbia*, Beggs and Mark forged a curious double act. Neither of them really trusted the other, but both were hugely energetic in pursuing a common goal: to persuade the new president of the United States that a space station should now be built. 'It is the logical next step,' Beggs stated.

Beggs argued that the Russians were still years ahead in terms of orbital experience, despite the American triumph of the 1969 moon landing, which was already getting to be 'yesterday's glory'. Meanwhile, Mark tried to persuade his contacts within the Air Force, and the DoD in general, to take an interest in the station. He also suggested to the Central

**BEHIND SCHEDULE**
An early version of the space station design, derived from a speculative study in 1982 by the Rockwell company (builder of the space shuttle orbiters). NASA had hoped that this station, or one very like it, would be in orbit by the early 1990s.

Intelligence Agency (the CIA) that such a large orbiting platform might be suitable for spying. He was trying to capitalize on Reagan's current enthusiasm for the strategic benefits of space flight. But the CIA's satellite experts were adamant that astronauts walking about inside the station would blur the surveillance cameras. Unmanned systems were cheaper as well as more efficient, they advised.

Beggs was not entirely happy at the idea of using his precious space station for military or spying activities. At first, he and Mark did not see eye to eye on the matter. Others within NASA were also anxious about military involvement. The entire issue simply evaporated when it became clear that the intelligence and defence establishments were all bitterly opposed to the station anyway.

It is worth noting that the 'black' budget for top-secret spy satellites has always been immense. These projects are called 'black' because everybody—even Congress—is kept completely in the dark about their true nature and their costs. In the late 1970s, the budgets for CIA and certain secret defence-related space projects actually *exceeded* civilian spending. NASA's space station represented an unwelcome competitor for funds. Meanwhile, America's total military budgets average $250 billion a year in comparison to NASA's $17 billion.

Conflicts of interest like this between NASA, the DoD, the CIA and many other large federal agencies required an elaborate bureaucratic system of buffering. Within the upper echelons of government, a body called the Senior Interagency Group (SIG) had been created to try and iron out such problems. In October 1982 a special Working Group was convened to deal specifically with space station issues. John Hodge, a British-born veteran of the earliest Mercury, Gemini and Apollo space projects, chaired the meetings with flair and determination, but he could not get the various representatives to agree on a common policy. Hodge had worked hard within NASA, planning a space station management structure, looking at the technical problems, and so forth. Persuading people *outside* NASA to conform was a tougher challenge

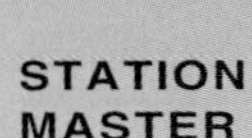

**STATION MASTER**
James Beggs, NASA's chief from 1981 to 1985, championed the space station effort, gained support from President Reagan, and encouraged international collaboration.

Meanwhile, Caspar Weinberger, who had done so much to save NASA's shuttle program from Nixon's cost cutting in 1971, was unwilling to be so helpful this time. He flatly opposed the space station. Weinberger was now Reagan's secretary of defence, and he was anxious to preserve as many shuttle flights as he could for defence-related activities. He believed the proposed station would swallow up launch capacity and deny his department valuable access to space. Laser satellites and missile platforms were more important than NASA's toys. The DoD representatives in Hodge's Working Group made their objections all too clear. The Group created a number of sample proposals but could not reach agreement about which of these proposals, if any, should be submitted to President Reagan for final approval.

## BREAKING THE LOGJAM

Thwarted by Weinberger, by the CIA 'spooks', by hostile White House staffers and by the endless committees and referral processes, James Beggs enlisted the help of a sympathetic White House insider, Colonel Gil Rye, to gain direct personal access to President Reagan. Rye was an official for the National Security Council, a very high-level group of cabinet advisors. In contrast to the DoD's rather narrow, parochial approach, the Security Council was able to take a wider view of American interests—a view in which a space station was seen as a beneficial tool of geopolitics and economic superiority. Reagan's National Security chief, Robert McFarlane, and his deputy, Admiral John Poindexter, counter-balanced Weinberger's opposition to the station.

On April 7, 1983, Rye arranged a meeting between Beggs and the president. To Beggs's relief, Reagan turned out to be broadly sympathetic to NASA's latest plan. After all, in his previous political incarnation Reagan had been governor of California, a state containing some very large aerospace contractors—most notably, Rockwell, builders of the space shuttle. Quite apart from the domestic political resonances, Reagan believed that a space station would fit neatly into his vision of a strong America. But many people within his administration did not approve. Reagan did not like to make important decisions without the consensus of all his senior staff, and his ultimate decision remained in doubt.

By August 1983, Hodge's SIG Working Group had thrashed out a space station proposal for Reagan, but certain members of the Group, fiercely hostile to the last, refused to allow the briefing papers to be submitted for the president's inspection. Colonel Rye may have been instrumental in breaking the logjam by going behind the backs of the station's opponents. By the end of 1983, he realized that SIG's main problem centered around the traditional political battle between NASA and the Defense establishment. It was time to bring in a third party, the Trade and Commerce Council, representatives of all those people within the country *apart* from NASA who would most stand to benefit from a space station program: the half-dozen or so giant and politically powerful aerospace corporations; the hundreds of smaller subcontractors; the tens of thousands of factory workers all across the country; the congressmen and women who represented them and counted on their votes—and on the election campaign funds so kindly donated by the big corporations to politicians who were in sympathy with them... In the end, the station's opponents at the DoD would prove no match against the mighty dollar.

On December 1, 1983, Rye brought the Commerce people into the White House for a discreet meeting with Reagan. With commercial interests now persuaded over to NASA's side, Reagan used this session as a cue for approval. He announced the space station project in his State of the Union Address on January 25, 1984, as high-profile an occasion as Beggs could have hoped for. 'Tonight, I am directing NASA to develop a permanently manned space station, and to do it within a decade... We want our friends to help us meet this challenge and share in the benefits. NASA will invite other countries to participate, so we can strengthen peace, build prosperity and expand freedom for all who share our goals.'

## MAKING ENEMIES

Beggs, Mark and Rye had achieved a significant milestone: presidential approval for a big new manned space project. In the process they had upset some very powerful people. Reagan's principal science advisor, George Keyworth, had come to the White House with a background in atomic weapons research. He was not even remotely supportive towards NASA and their spaceships. Worst of all, Keyworth had been entirely bypassed by Rye's tactics, and was furious that Beggs had shown space station plans to Reagan without his knowledge or approval. The SDI lobby within the White House were largely on Keyworth's side. They were concerned that a NASA station would take away resources from their own pet schemes.

Pitched battles between presidential science advisors and NASA chiefs are not uncommon. In the early 1960s, President Kennedy's science advisor, Jerome Weisner, had vigorously opposed Jim Webb's plans to go to the moon. On one notable occasion, a British government official on a diplomatic visit witnessed a stand-up row between Weisner and Webb. Later, and in private, the British visitor asked President Kennedy, 'Who will win?' Kennedy grinned and said, 'Webb will win. He's got all the money, and Jerry Weisner only has me.' Twenty years later, NASA's chief would find the shoe was on the other foot. Keyworth now had all the money, and Beggs had only the president.

Keyworth was not a safe person to offend. In fact, the Reagan White House as a whole was never a safe place to make enemies. Within the next couple of years, Rye would quietly leave to pursue a private career. Hans Mark, rebuffed by his 'friends' in the SDI establishment, would resign as Begg's deputy to take up the chancellorship of Texas University. And Beggs, stubbornly unwilling to acknowledge the danger presented by some of his Washington opponents, would soon have his entire career and reputation at NASA thrown into doubt, possibly as the result of political malevolence directed towards him.

The station itself, and the back-door approval process engineered by Rye, were by no means the only matters to upset the apple cart. President Reagan's speech had contained yet another surprise for NASA's enemies.

## THE INTERNATIONAL BLUFF

Shortly before his State of the Union speech, Reagan had alerted European, Japanese and Canadian leaders of his intentions to invite their participation, sending them private memos. His public announcement on January 25 came as a surprise to many people in America itself. How was it that Reagan could have made such a visible commitment to foreign cooperation without anybody knowing in advance?

In conjunction with just a handful of sympathetic senior political figures, James Beggs's team (and especially NASA's hardworking experts on international affairs, Kenneth Pedersen and Peggy Finarelli) had managed to incorporate an international element into Reagan's speech, with very little external consultation. In other words, they hadn't

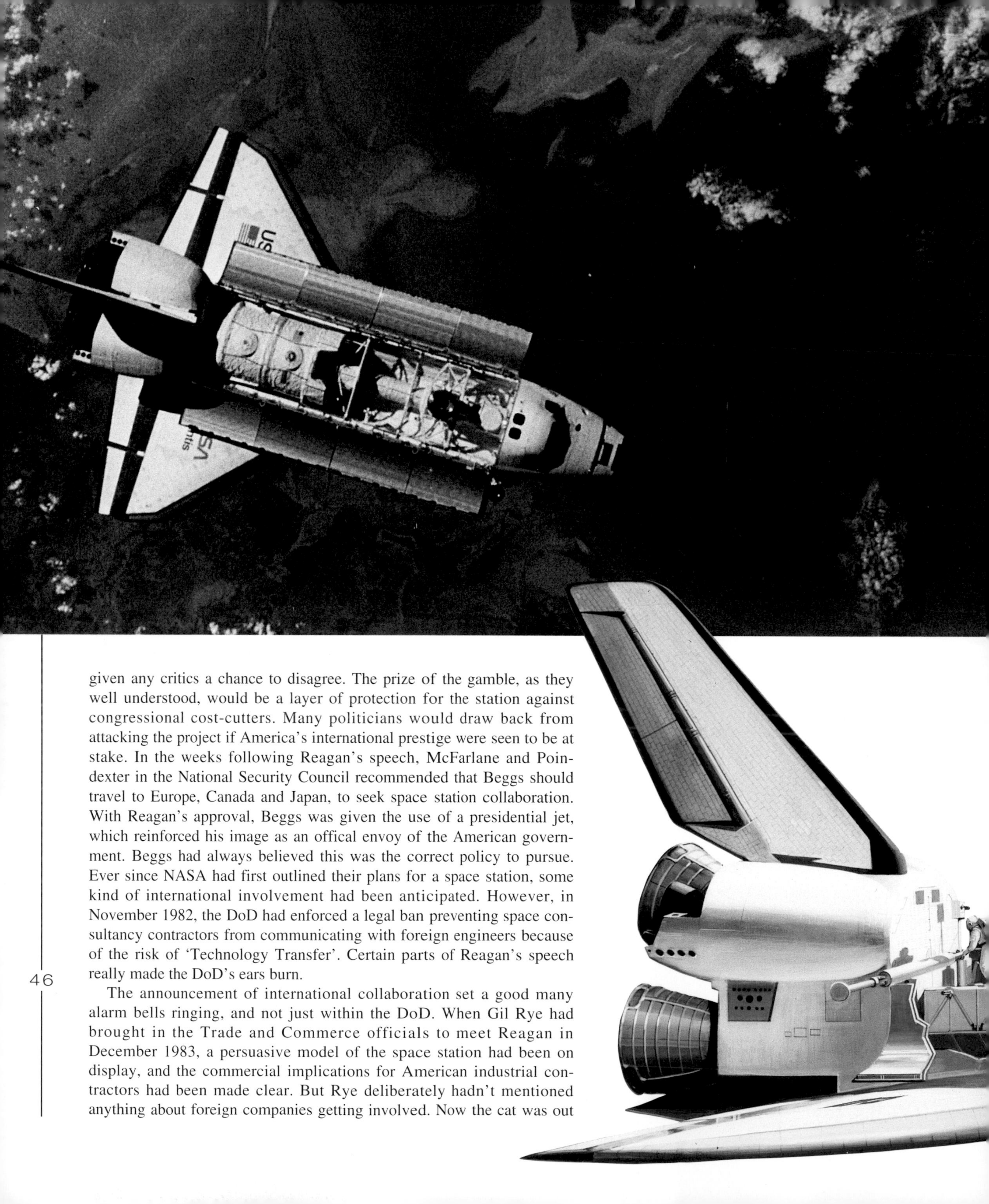

given any critics a chance to disagree. The prize of the gamble, as they well understood, would be a layer of protection for the station against congressional cost-cutters. Many politicians would draw back from attacking the project if America's international prestige were seen to be at stake. In the weeks following Reagan's speech, McFarlane and Poindexter in the National Security Council recommended that Beggs should travel to Europe, Canada and Japan, to seek space station collaboration. With Reagan's approval, Beggs was given the use of a presidential jet, which reinforced his image as an offical envoy of the American government. Beggs had always believed this was the correct policy to pursue. Ever since NASA had first outlined their plans for a space station, some kind of international involvement had been anticipated. However, in November 1982, the DoD had enforced a legal ban preventing space consultancy contractors from communicating with foreign engineers because of the risk of 'Technology Transfer'. Certain parts of Reagan's speech really made the DoD's ears burn.

The announcement of international collaboration set a good many alarm bells ringing, and not just within the DoD. When Gil Rye had brought in the Trade and Commerce officials to meet Reagan in December 1983, a persuasive model of the space station had been on display, and the commercial implications for American industrial contractors had been made clear. But Rye deliberately hadn't mentioned anything about foreign companies getting involved. Now the cat was out of the bag, and Corporate America was angry. NASA's own field centers were also concerned. How much control would be sacrificed to foreign agencies? Who would be responsible for the extra layers of management?

In response, Beggs asked his European counterparts if they could perhaps focus on broad issues, and on how they might use a future space station, rather than on how it might actually be built. But the Europeans made it clear that they wanted to be involved in construction. They had their own aerospace companies to worry about, and they would want to place hardware contracts within their own borders. In many ways, Beggs's international gamble looked like it might create more problems than it would solve.

But Beggs had one more card up his sleeve. In fact, ever since he'd first joined NASA, he'd been dealing from the bottom of the deck as far as the international problem was concerned. In 1982 he travelled widely, seeking purely informal agreements on space station studies from other countries. During the summer of 1983, he convened a symposium of interested parties from Europe, Canada and Japan, along with American members of Congress, and representatives from the DoD and other major government departments. In his keynote address Beggs stated that the purpose of a space station 'is, of course, to maintain our leadership'. The foreign delegates' discomfort at the thought of playing second-fiddle to NASA turned to outright amazement when Beggs went on to say, 'If we can attract international cooperation, then other nations will be co-operating with us in the resources that they spend on space, rather than competing with us.'

**EUROPE IN SPACE**
June 1995: the space shuttle *Atlantis* in orbit above the Aral Sea, with a European-built Spacelab in the cargo bay. The artwork below shows a typical Spacelab configuration as carried in a shuttle. Different combinations of hardware can be flown, allowing extra capacity for pressurized human physiology experiments in the cylindrical compartment, or else giving preference to experiments in the exposed pallet racks. The pressure module is 4m (13.5ft) in diameter. A typical Spacelab weighs 11,350kg (25,000lb).

Of course, Beggs's speech was intended primarily to reassure the American delegates within the symposium. His implication was clear. Bringing foreign construction companies into the space station project would partially divert those countries from building independent competitive capabilities in space, thus ultimately saving American jobs, not threatening them. The other prospective space station 'partners' were amazed that Beggs could be so blatant about American self-interest.

Beggs's speech may have been biased towards his home audience, but it did contain an essential truth. Joining the NASA space station effort would indeed take away resources from independent foreign projects, unless they could come up with much more money than they were already spending. But President Reagan's invitation was difficult to turn down without giving the appearance of lacking technological courage or know-how. Despite their doubts, ESA eventually decided that the station might be a suitable objective to pursue. Of course, 'space cadets' in all the member countries pushed their governments just as hard as NASA had pushed the White House.

## ACROSS THE ATLANTIC

Europe's space effort, with French, German, Italian and so many other vested national interests involved, might well have collapsed by now, had it not been for a continent-wide decision more than two decades ago to create a common voice for their ambitions: the European Space Agency (ESA). Founded in 1973, ESA shared many of the same objectives as NASA: to pull disparate efforts together into a single cohesive program. The American agency, and their Russian counterparts, have had to acknowledge them as a third significant player in space.

An existing French experimental rocket facility in Korou, French Guiana, was expanded as a launch center for a new generation of European rockets. On Christmas Eve, 1979, ESA's powerful 50m-long (160ft) 'Ariane' booster made its first flight. In the following year, a private European business consortium, Arianespace, was established to build and fly these rockets for profit. The consortium was underwritten by 36 major aerospace and technology companies from around Europe, backed by 13 major banks. This was a very ambitious commercial proposition, presenting a major threat to NASA's waning supremacy in the unmanned launch business. Despite occasional failures, Ariane proved to be a good vehicle, capable of reaching the high orbits required for communications and earth-observation satellites. By May 22, 1984, Arianespace was ready to fly its first customer into orbit—an *American* commercial satellite.

NASA have always regarded ESA as a threat and as a potential partner. America has courted broad European involvement with space activities since 1969. As soon as ESA became operational in 1973, NASA asked them formally to collaborate on 'Spacelab', a pressurized scientific module to be carried in the cargo bay of an American space shuttle—at least, once the shuttle was built. During the 1970s, at the height of their efforts to sell the shuttle concept to Congress, NASA knew that some kind of European contribution would compensate for the fact that they couldn't yet afford both a shuttle and a space station. In fact, Spacelab had originally been just one of three collaborative options offered to Europe. Even as the ESA structure was beginning to formulate in the early 1970s, NASA had invited them to help develop parts of the shuttle itself, in particular the tail section and the 18m-long (60ft) payload bay doors. A second option was to design an intermediate booster for satellites and small 'tug' vehicles, which could be used as part of an integrated shuttle transportation system. These were attractive options, from which NASA pulled back after 1972 when the DoD came up with their favourite objection: Technology Transfer. Security-minded officials within the American government did not want the shuttles to become reliant on any foreign hardware. The European community spent $20 million researching potential work on the shuttle before NASA politely told them, in 1973, to accept Spacelab or nothing at all. Spacelab was a guest payload aboard the shuttle that did not require sensitive shuttle hardware details to be revealed during its construction.

**SETTING A PRECEDENT** the Spacelab experience proved valuable when it came to designing potential interior working areas for the new space station, as depicted in this 1989 NASA artwork.

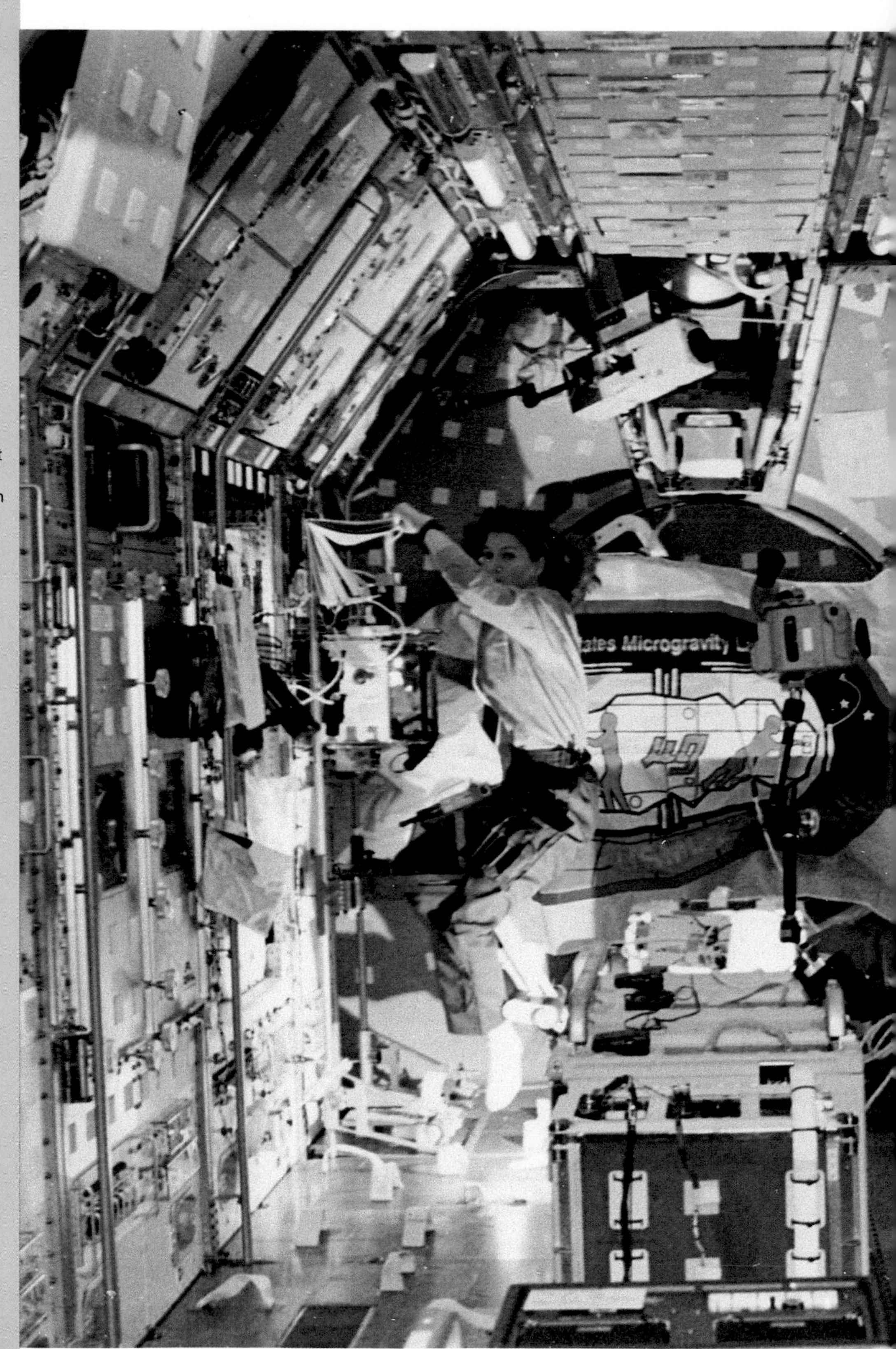


**INSIDE SPACELAB**
A typical interior view of Europe's principal manned space asset. In 1975 NASA collaborated with ESA to build Spacelab, a scientific work module carried in the payload bay of a shuttle. The lab's interior is a tidy and well-organized working environment, with equipment racks that can be switched for a range of mission requirements. A version of the lab flew for the first time on November 28, 1983. John Young commanded the shuttle *Columbia*, with co-pilot Brewster Shaw. The American Spacelab crew members were Robert Parker and Owen Garriott (a Skylab veteran). Europe's crew were Byron Lichtenberg and Ulf Merbold.

## SPACELAB

Having spent $20 million for nothing, the European space community took an expensive bruising from their unpredictable American counterparts. But still, they swallowed their pride and made a new deal. A Memorandum of Agreement between NASA and ESA was signed on August 14, 1973. The plan was that European industry should provide a complete Spacelab module free of charge, with Germany taking the lead role in design, funding and construction; and in return, American shuttles would fly the lab, along with European astronauts. NASA would then purchase from ESA the hardware for 'at least one' set of Spacelab modules, and perhaps as many as half a dozen, thus helping to defer European development costs.

The first Spacelab was built for a cost equivalent to $1 billion; more than twice the anticipated budget. Certainly ESA had tried to keep their side of the bargain, but NASA's development of the shuttle took longer than expected. Even though Spacelab was a guest payload and not part of the shuttle itself, it was designed to fit very snugly into the cargo bay. Inevitably, redesigns of American hardware generated expensive changes in ESA's systems. But construction work progressed without too many upsets, and 'Spacelab 1' eventually flew for the first time on November 28, 1983, in the back of the shuttle *Columbia*, tended by Dr Ulf Merbold, the first European astronaut. Instrument problems delayed the next mission, but by April 1985, Spacelab was up again, this time aboard NASA's shuttle *Challenger*.

Spacelab is a modular system of interchangeable pressurized compartments and external 'pallets' carrying experiments that need raw exposure to space. The pressurized crew sections are a fraction over 4m (13ft) in diameter, which is the widest load a shuttle's cargo bay can carry. A narrow connecting tunnel links the lab with the shuttle's flight deck, and the modules are crammed with computers and control equipment for the external experiments. All in all, Spacelab is a fine system. During its first flights, the primary limitation on its peformance was that the host shuttles could not remain in orbit for much longer than a week.

During 1984, after Beggs and Reagan had first reached agreement on the space station, the NASA chief travelled around Europe, encouraging ESA's member nations to come on board. But at first, ESA strategists were concerned that the project might make their new Spacelab redundant just as it was starting to prove its value. NASA assured them that the station was a long-term option, and Spacelab would still have a useful intermediate life. Gradually, and then with increasing enthusiasm, ESA came aboard, unwilling to be left behind. But after their previous experiences working with America—and being well aware of NASA's desire to 'maintain leadership' in the station project—ESA hoped eventually to reduce their future dependence on American hardware. Spacelab wasn't being flown quite as often as they had been promised (10 flights in 15 years) because the shuttle itself wasn't delivering the anticipated 60-a-year launch rate. Total shuttle flights averaged barely six a year, and American payloads were still dominating when they did fly.

Worst of all, NASA's original promise to buy 'at least one' complete spacelab from ESA, with a hint of maybe another half a dozen to come, resulted in just the one firm order specified in their contract as the legal minimum purchase. NASA needed all their spare cash for the shuttle, and ESA were thus unable to recover the costs of building Spacelab by selling extra equipment. Next time, ESA decided, they would make tougher contracts with no loose ends. It was a lesson in what the Americans would call 'hardball' politics.

## SEEKING INDEPENDENCE

The tensions between America and Europe were redeemed by some important considerations that had nothing to do with money or politics. Ground-floor engineers and scientists on both sides revelled in their joint efforts. These nuts-and-bolts people really did speak a common language, and something positive had emerged from the Spacelab experience. NASA liked to imagine that they had taught the Europeans how to run a large-scale space project. ESA managers weren't too sure about that, but they had won a solid foothold in manned spaceflight, conducting a valuable range of scientific investigations in the process.

Working with NASA was a great deal better than having no manned access to space at all, but by 1985, ESA's desire for some degree of independence had hardened into a formal process, goodwill or no. They began detailed engineering work for their own mini-shuttle, the 'Hermes', along with a more powerful version of their Ariane rocket, to boost Hermes and other large payloads into orbit. ESA also made it clear they would take second place to nobody. On January 31, 1985, they prepared a strongly worded official statement of their position with regard to the space station: 'Europe will seek appropriate participation, with access and use of all elements of the station on a non-discriminatory basis.' ESA would not accept NASA as the unchallenged 'leader' of the project. Three years of negotiations with the Americans now followed.

By 1987 the central element of ESA's plans had become a project called 'Columbus', and without this program, their Hermes spaceplane was pointless. Columbus consisted of three distinct elements: a small unmanned 'free-flying' experiments platform that would drift in space under automatic control, occasionally refurbished by Hermes; a larger man-tended free-flyer capable of docking with the Hermes spaceplane and serving as a small-scale space station for short periods at a time; and most important of all, a big pressurized module designed for attachment to the NASA space station. Despite all their tough words, ESA's ambitions still centered heavily around America's promises.

## FREEDOM IN CHAINS

When President Reagan announced the space station project in 1984, a development program of eight years was envisaged, along with a budget of $8 billion over that period. James Beggs was delighted, feeling that he'd provided NASA with a really ambitious project to get their teeth into, but some of his senior staffers were not so pleased. For Project Apollo's veterans, an earth-orbiting station wasn't nearly enough. Billions were being spent on defence and SDI, and the people who had once conquered the moon believed the agency should have fought for a larger share of spending for a manned mission to Mars.

As it turned out, Beggs's achievement was a substantial one. The space station was a much grander and more expensive proposition than NASA had initially indicated. Six months after Reagan's announcement, congressional budget analysts took a closer look at the station's likely costs. The nominal price of $8 billion took into account the fabric of the hardware itself, but not its long-term operating costs, nor all the costs of flights for launching the separate components. By the time the various associated budgets were factored in, the price tag threatened to top $20 billion. The fact that politicians on Capitol Hill continued to take any further interest after the true cost became apparent is a tribute to Beggs's tenacity. Reagan was a popular president, and when Beggs persuaded him to support the space station, NASA gained valuable political momentum. Beggs's old colleague and deputy, Hans Mark, has said of him, 'He will be remembered as a NASA administrator who succeeded in persuading a president to do something new.'

Unfortunately, American space policy is seldom decided by presidents alone. NASA's budget requests are reviewed on a yearly basis by a series of congressional committees. The station's budget was approved piecemeal, a few hundred million dollars a year, escalating by the late 1980s to around $1.5 billion a year, but Congress kept pulling back from any substantial long-term commitments. A dreary battle, self-defeatingly expensive, and extending over more than a decade, was joined between NASA and the politicians on Capitol Hill for the future of the space station. The most controversial element of the debate was at the same

**RISKY BUSINESS**
Astronauts strapped in their couches in the mid-deck of a shuttle. The crew compartment can accommodate seven people, three in the mid-deck and four on the flight deck. All crew members wear pressure suits during launch in case of an accident. The shuttle launch phase is very risky. Two solid rocket boosters assist the first three minutes of flight, and once lit they cannot be turned off. A shuttle cannot discard them until their fuel content is completely exhausted. This design feature appalled many veteran Apollo engineers who could not understand how NASA could build a vehicle with no launch escape facility. All previous capsules had ejection systems or compact rocket thrusters to blast the crew compartments away from faulty boosters automatically at a moment's notice.

time the most obvious one. Nobody, and least of all NASA's keenest supporters, could explain what a space station's ultimate purpose would be. This problem was heightened by disagreements within NASA.

Although the agency is commonly perceived as a single body, this has never been the case. In 1958, it was created mainly from an amalgam of existing research centers in rocketry, aircraft testing, astronomy, science and electronics, all with their own long histories and traditions. These became the so-called 'Field Centers', the widely distributed complexes that actually create and run the space programs on a day-to-day level, with each center supposedly specializing in a different activity. During the 1960s, strong central management from NASA's Washington headquarters glued this loose alliance into a cohesive whole. Webb and his most senior associates decided what NASA should do. The field centers

then determined how to carry out the missions assigned to them. But after the Apollo era, funds for space activities were harder to obtain, and NASA chiefs had to spend more time on creating a strategy for survival, not in the depths of space, but in the Washington political arena. Field centers were gradually entrusted with more responsibility for determining NASA's future space strategy, so that headquarters could concentrate on dealing with Congress and the White House. Devolution of power to the field centers may have been the space agency's single greatest management error, from which so many of its more public ills have stemmed.

**FLYING BOMB**
Fuelled for lift-off, a typical shuttle carries the explosive potential of a small nuclear bomb. The launch pad is sited several kilometres away from any inhabited buildings. The low structures near the pad are either unmanned machine facilities, or else reinforced concrete bunkers, occupied only before launch.

## INTERNECINE WARFARE

In 1984 it was not enough for James Beggs to seek political, industrial and international consensus for his space station. NASA now contained individual constituencies, which also had to be pacified—the field centers had become like rival barons in a feudal court. Three of them in particular are familiar to the general public: the Kennedy Center in Florida, which handles launch operations, the Marshall Center in Alabama, which develops the major booster systems and other large structures, and the Johnson Center in Texas, which runs the manned missions, trains the astronauts and develops the caspules and orbiters. Beggs wanted the support of all the centers, and as a consequence he initially avoided centralizing the station program at any one of them. In particular, he had to worry about the long-standing rivalry between Marshall and Johnson.

John Hodge, who had run a Space Station Task Force out of Washington, had pulled together a close-knit team of representatives from all the centers, just as in the Apollo days. Their conceptual work had contributed greatly to getting presidential approval. Now that it was time to start handing out real hardware contracts and assigning real, as opposed to theoretical field center responsibilities, Hodge was determined that some kind of central control should be maintained. But his influence waned, and the various center directors started sharing out portions of the station project among themselves, and over Hodge's head—and even, to some extent, over Beggs's. The contractor companies also competed fiercely for potential work, forging separate and sometimes cluttered alliances with rival centers. What emerged was a concept called the 'Work Package' in which each of the major centers would be assigned portions of the station to develop. The centers would manage their own teams of contractors. But defining which portion belonged to what center led to many difficulties, because the space station would have to be a complex assembly of related parts. Marshall's experience developing Saturn V and Skylab under Wernher von Braun led them to take on responsibility for 'structures' like the station's pressurized habitation and lab modules. As general space program managers, Johnson was given responsibility for the station's design as a whole, including the lattice metal trussbeam elements that held everything together. But since the trusses were structural, why wasn't Marshall handling that? A further complication arose out of the different philosophies at the centers. Marshall wanted a gradual construction of the station, starting with the truss beams and adding modules and manned capability over time. This was known as the 'outside-in' approach. Johnson wanted the opposite: manned capability as soon as possible, starting with the modules and adding trusses later. Naturally this was called 'inside-out' construction.

## CLASHING CULTURES

Marshall and Johnson couldn't even agree what the station was for. Johnson, with their experience of controlling moon flights and shuttle missions, wanted to use it as a base for continued exploration of space—a classic 'staging post' alongside which future lunar and interplanetary ships could be built in orbit. Even when this notion became unrealistic because of NASA's general budget restraints, Johnson wanted at least to retain this capability as a long-term option. But Marshall, with Skylab and their collaboration on ESA's Spacelab on their list of achievements, favoured a more conservative scientific and experiments-based platform.



The field centers were becoming increasingly ungovernable. After so many years of providing jobs and contracts in their respective states, they had secured considerable bargaining power with their local congressional representatives. This was corrosive to NASA as a whole. For instance, Washington headquarters might instruct Marshall to make a program reduction, only to be confronted by a delegation of Alabaman politicians threatening to generate a fuss in Congress on Marshall's behalf; while

Johnson always benefited from the traditional strong political and financial influence of Texan lawmakers in Congress. These field center tactics seemed undignified to many outsiders, who still tended to see NASA as the more cohesive entity of the Apollo years. The space program as a whole gained a new and unwelcome reputation as fractious and poorly managed. Beggs was a strong and capable administrator, but he was too focused on achieving consensus. What was needed right now was some bloody-minded ruthlessness. But these problems were trivial compared to what was to come.

## THE *CHALLENGER* EXPLOSION

Winter of 1985 saw the American space agency at a low ebb. Hans Mark had left NASA and James Beggs's new deputy was Bill Graham, a White House political appointee sponsored by Reagan's chief science advisor, George Keyworth. A legal problem stemming from Beggs's previous commercial career forced him to step aside from NASA's top job. Graham became acting administrator, but Beggs continued to run an office from the agency's headquarters. For a while, a gruesome situation developed during which Beggs and Graham appeared to be running rival fiefdoms within the corridors of NASA. Beggs felt betrayed by the White House, and wondered if his legal problems had been artificially exaggerated by political schemers so as to lever him out of NASA in Graham's favour. (Eventually, every charge against him was dropped.) This particular feud gives some flavour of the many tensions preoccupying senior NASA managers at the time.

Graham was not a rocket expert, but his management instincts were sound enough. Towards the end of November 1985, he noticed some odd technical discrepancies in documents analysing the reliability of shuttle rocket motors. Genuinely disturbed, he made a note requesting better data. But information flow within the agency was log-jammed at all levels. Middle-ranking managers found their allegiances shifting this way and that between Graham and Beggs: a circumstance which generated dangerous confusions. Morale at all levels was dampened, not just by the leadership crisis, but by the increasing pressures of the shuttle schedule.

On January 28, at 11:38 a.m., *Challenger* lifted off. Seventy-three seconds into the flight, it fell out of the sky in a shower of burning fragments. All seven crew members were killed: Commander Dick Scobee and his co-pilot Mike Smith; mission specialists Judy Resnick, Ellison Onizuka, Ron McNair and Greg Jarvis; and a high school teacher launched as part of a schools' education program, Christa McAuliffe. Shocked by the tragedy, President Reagan ordered an immediate inquiry, the Rogers Commission, to investigate the causes of *Challenger*'s loss.



It emerged that hardware failures were only symptoms of the disaster, and not the ultimate cause. The Rogers Commission investigated the fraught relationships between NASA's field centers. The complications were not just a space station issue. They were endemic. The physicist Richard Feynman sat as a member of the Commission. He rightly identified that NASA had lost control not only of themselves but also of their

**THE SOLID BOOSTER PROBLEM**
The photo to the left shows a typical shuttle launch dominated by the output of the twin solid rocket boosters. The segmented construction of a booster can be seen. To the right, the space shuttle *Challenger* is destroyed on the morning of January 28, 1986. The central ball of flame and smoke is the liquid fuel tank exploding. The smaller plume shooting ahead of the main cloud of debris (*top right*) is the exhaust trail of a solid rocket booster, blown off the main body of the ship and still firing. The technical explanation turned out to be straightforward. A rubber seal between two mated sections of the starboard solid rocket booster had failed to seat itself properly because icy weather had made the rubber stiff and unpliable. A small gap in the seal opened up, and hot combustion gas leaked out of the side of the booster, acting like a rogue thruster and tilting the ship slightly off-course. Between the two boosters, and slung underneath the winged orbiter, there was a huge tank containing liquid propellants for the orbiter's three engines. The escaping flames scorched the side of this tank like a welder's torch, and the vehicle exploded.

contractors, the myriad private companies who fabricated all the shuttle hardware and support facilities in return for large sums of government money. James Webb's legacy of distributing the Apollo-era benefits of space had become a mess. Feynman noted, 'A NASA center in Huntsville designs the engines, the Rocketdyne company builds them, the Lockheed company writes the instructions, and then the Kennedy launch center installs them! It may be a genius system of organization, but it seems a complete fuzzdazzle to me.'

In the wake of the *Challenger* disaster, the space station's design work carried on, but many politicians within Congress were now convinced that NASA was incapable of completing it. By 1988 the main advance had been giving it a name, 'Freedom'. NASA embarked on a series of re-designs, scaling down their ambitions, but far from emerging as a leaner, cheaper system, Freedom's anticipated budget simply climbed upwards.

By 1992, it threatened to cost $31 billion. Not least of the problems was factoring in the $8 billion NASA had already spent on trying to make it cheaper. So far, the effort to build Freedom had been, as Richard Feynman might have said, a complete fuzzdazzle.

In addition to these setbacks, the long-awaited Hubble space telescope, launched by the shuttle *Discovery* on April 24, 1990, beamed back images that were out of focus.The $2 billion orbiting Newtonian reflecting telescope, was a decade in the planning. Astronomers around the world had been looking forward to putting instruments out in space, beyond the distortions of the earth's atmosphere. The silly and avoidable miscalculation of the mirror's optics during manufacture crippled Hubble's performance at a stroke. The press had reported *Challenger*'s destruction in 1986 as a national tragedy. They portrayed Hubble as farce. Reagan's successor George Bush and his Vice-President Dan Quayle now regarded NASA as a severe liability to their political standing.

## PUTTING FREEDOM TO THE VOTE

The White House appointed Norman Augustine to head an independent inquiry into NASA's affairs. As chief executive of the Martin Marietta company, he was widely respected as a capable and honest man in an industry renowned for its wastefulness and sharp practices. His probity and reputation made him a hard man to ignore. 'If you want bad news broken gently, Augustine's your man,' the London *Economist* said at the time. And he was gentle, because the news *was* bad. The 'Augustine Report' of December 1990 stated that NASA had lost their way. The shuttles and the Freedom project were soaking up funds needed for other work: 'Space science projects deserve funding over and above space stations, spaceplanes, manned missions to the planets and other major pursuits which often receive greater visibility.' Augustine also stated that NASA's technology was badly outdated: 'America has not initiated development of a new rocket engine—the muscle of any space pursuit—in nearly two decades.' The use of the shuttle as an all-purpose workhorse also came under fire: 'In hindsight it was inappropriate in the case of *Challenger* to risk the lives of seven astronauts and one-fourth of NASA's shuttle launch assets simply to orbit a communications satellite.'

Most damaging of all for the space station's supporters, Augustine judged that it was too big, too complicated and too expensive. A smaller and cheaper redesign was recommended. Sniffing the political wind, Congress anticipated Augustine's analysis by some weeks. The space agency had hoped for close to $2.5 billion to be granted for their scheduled development work on the station during the coming year. Congress slashed $500 million from that request, warning NASA that another $600 million could go if the redesign was inadequate.

On June 6, 1991 the space station came under intense scrutiny within Congress. A small but powerful Appropriations Committee had already voted not to assign NASA any money that year for the station; a decision that forced a major vote in the 435-strong House of Representatives. Perhaps surprisingly, this was the first time since 1984 that Congress had

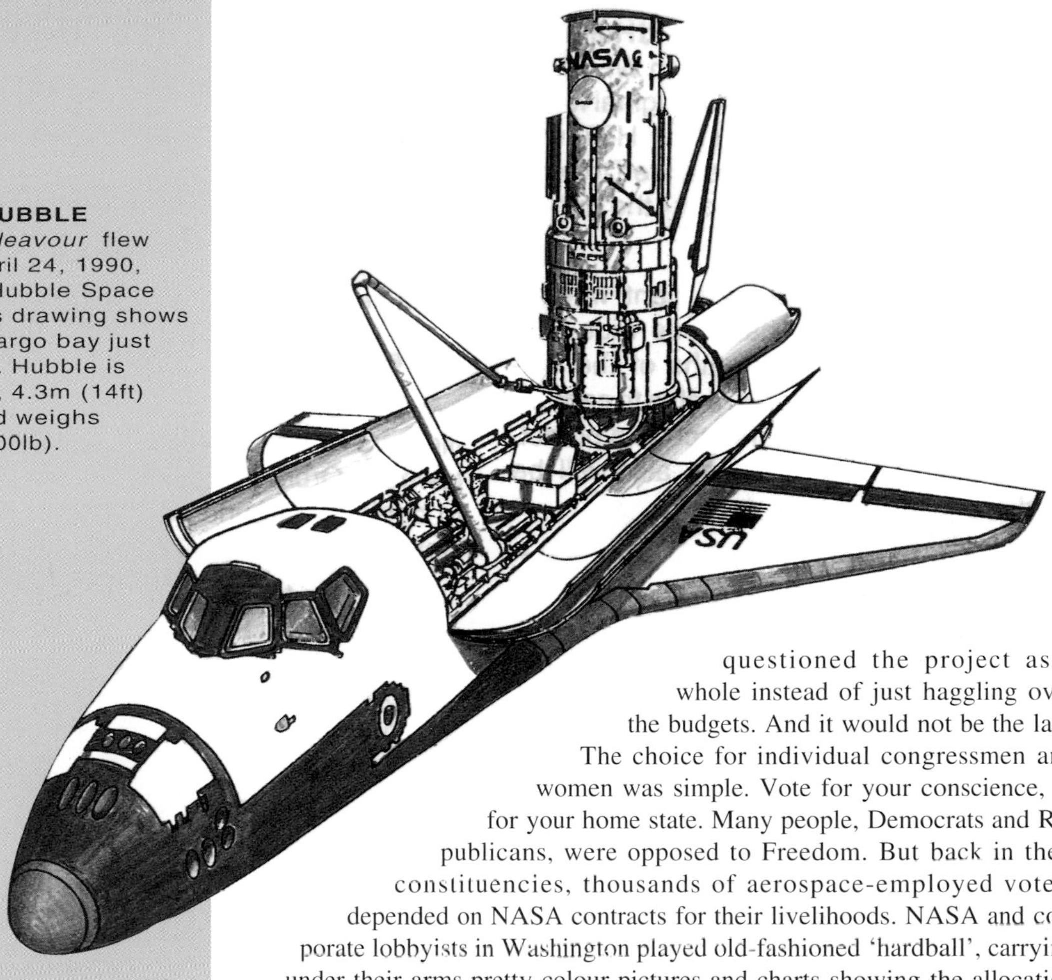

**LAUNCHING HUBBLE**
The shuttle *Endeavour* flew into orbit on April 24, 1990, to release the Hubble Space Telescope. This drawing shows Hubble in the cargo bay just prior to release. Hubble is 13m (43ft) long, 4.3m (14ft) in diameter, and weighs 11,000kg (24,000lb).

questioned the project as a whole instead of just haggling over the budgets. And it would not be the last. The choice for individual congressmen and women was simple. Vote for your conscience, or for your home state. Many people, Democrats and Republicans, were opposed to Freedom. But back in their constituencies, thousands of aerospace-employed voters depended on NASA contracts for their livelihoods. NASA and corporate lobbyists in Washington played old-fashioned 'hardball', carrying under their arms pretty colour pictures and charts showing the allocation of space station jobs on a state-by-state basis.

James Beggs's 'international' gamble also began to pay some dividends at last. Several key congressional figures were now saying just what he had always hoped they would—if we cancel, then our allies will never trust us again. Europe, Canada and Japan had already spent $1.6 billion between them on Freedom, and they had indeed reacted with official expressions of anger at high diplomatic levels when they learned that Congress was questioning the project's future.

Freedom was supported off-stage by George Bush and Vice-President Dan Quayle, neither of whom wanted the humiliation of cancellation. It survived by 240 to 173 votes. NASA got the $2 billion they needed to keep on track for another year. But these funds relied partly on a pledge to draw money away from unmanned space activities. Despite Augustine's judgement that science projects should be protected against the glamorous but expensive manned systems, Freedom was becoming a parasite.

# FASTER, BETTER, CHEAPER?

*During the early 1990s the space station's design was overhauled in an effort to reduce its costs. NASA's new chief, Daniel Goldin, also imposed great changes within the agency, streamlining the levels of management and cutting jobs. Meanwhile, after the collapse of the Soviet Union, the American government encouraged NASA to integrate their station plans with the Russian space program. The reasons behind this were varied and complex.*

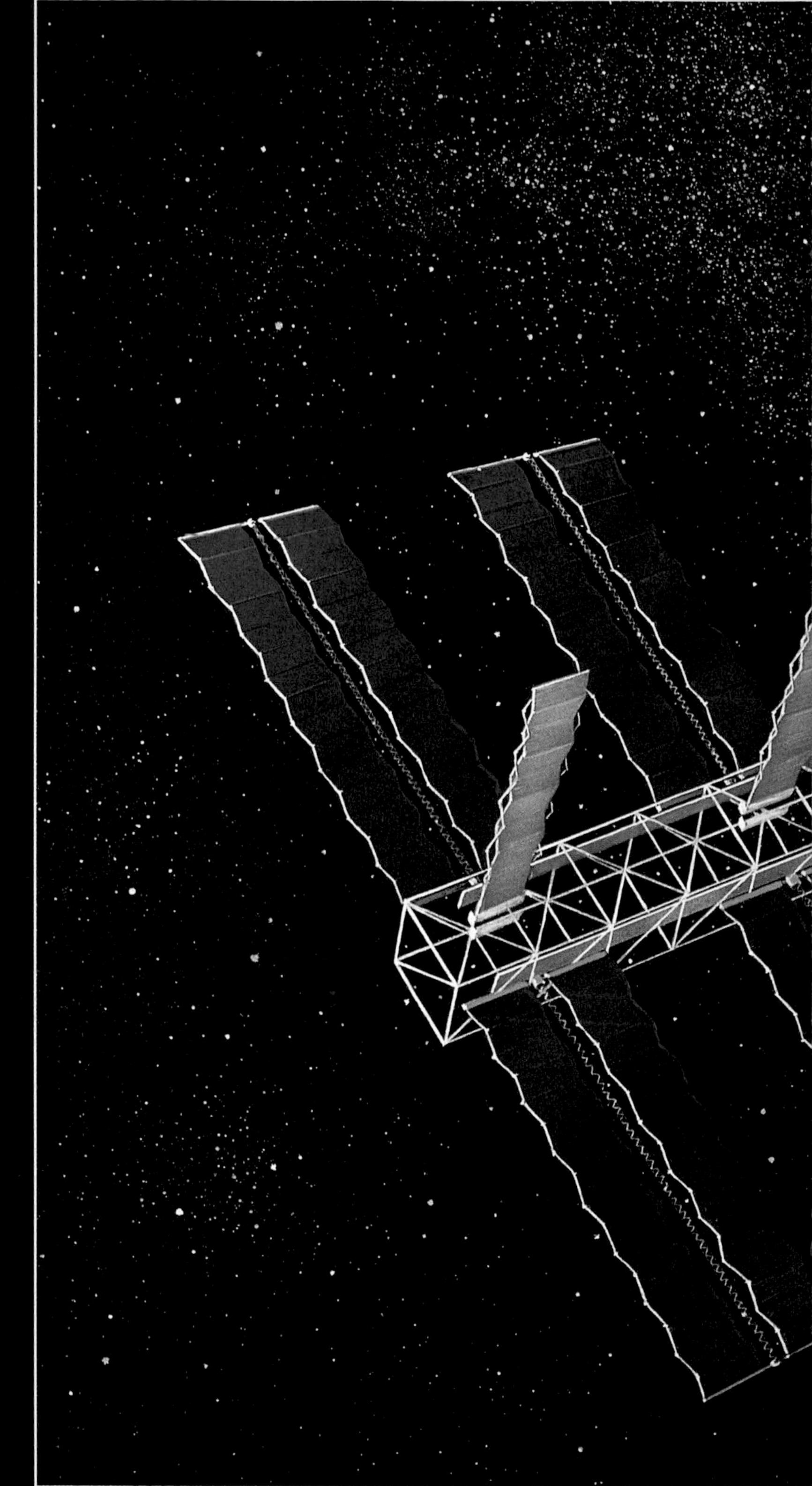

**LIMITED FREEDOM**
In this 1992 painting by David Hardy, the space station has shrunk, losing the dual-keel truss beams proposed in 1987.

HARDY

# FASTER, BETTER, CHEAPER?

Dan Goldin was raised in Brooklyn, New York by a working-class Jewish family. In May 1961 he was completing an engineering degree when he heard President John F. Kennedy pledge America to a moon landing. As soon as his degree was complete, Goldin rushed off to join NASA, and spent five years at their Lewis Research Center in Cleveland, Ohio. In 1967 he joined the TRW company in California for a 25-year career working on top-secret defence and satellite systems. He gained senior rank, but when George Bush invited him to take charge of NASA in 1992, Goldin was glad of the opportunity to run purely civilian space projects.

The first thing that became clear about his management style was his pugnacity. He didn't care about offending people. Jim Webb, Tom Paine, James Fletcher—all the previous NASA bosses had worked in the old Washington manner: all subtle politics and making the right connections over lunch at the club. Goldin's methodology was very different. He turned out to be a streetfighter with the instincts of his Brooklyn boyhood still very much intact. He favoured open combat.

His first enemies were the contractors. 'The corporations have had a great time at our expense for years, and now the party's over,' he once said, insisting that the contractors would never treat private customers the way they treated NASA. 'What they do to us is immoral!' At a luncheon in 1993, a journalist from *Air & Space* magazine witnessed Goldin getting into an argument with an executive from the powerful Hughes Satellite company. His voice rising as the roomful of guests looked on in shock, Goldin fumed, 'Your track record with NASA is terrible. You've overrun almost every contract you've had with us. It's unacceptable!'

Goldin was the most outspoken NASA boss in a generation. He fought the agency's corner like a lion. The trouble was, he savaged his own people as well as the contractors. 'This agency doesn't exist for the benefit of its employees. It exists to do a job for the nation.' Soon after his arrival at NASA in 1992, he backed up this philosophy with job cuts, sacking or replacing half the Field Center directors to reinforce his own authority from Washington headquarters. Then he shed 5,000 staff from what was once a 25,000-strong workforce. Thousands of contractor-related jobs also came under fire. And this turbulent treatment is not over yet. Goldin's current plans include 'outplacing' a further 5,000 NASA positions over the next four years, with contractors losing many more jobs on top of the thousands already gone.

**THE FINAL FREEDOM**
This 1991 artwork from the Rockwell company shows one of the last design iterations of Freedom before a program-wide change in 1993 brought the station close to its current (and final) configuration. The large rectangular array of truss beams proposed in 1987 has been eliminated, and the remaining central keel is scaled down from 150m (500ft) in length to 108m (355ft). By 1993 certain key components for Freedom were already at the initial construction stage. NASA claims that 75 per cent of these components have been successfully absorbed into their latest plans, thus avoiding waste.

Having sent NASA and its attendant manufacturing companies such a stern message about cost efficiencies, Goldin has, however, offered private industry a controversial big prize: the entire shuttle contract. Attempting to shave $500 million from the $3 billion shuttle-operating budget, he trimmed NASA's launch pad engineers, maintenance crews and managers. He then asked rival aerospace corporations to present a cheaper package for running the shuttles. After reconsidering the safety implications, Goldin offered the contract to the companies already most familiar with them: Rockwell, manufacturers of the winged orbiters, and Lockheed, providers of the launch staff. These two contractors have recently created a new company between them, the United Space Alliance. By 1998, shuttle maintenance, all launch preparations, and most of the associated activities within the mission control center will be almost entirely in their hands. 'I suppose if we ask nicely, they'll still let our astronauts fly occasionally,' one agency insider said. In February 1996, NASA's chief of shuttle operations, Bryan O'Connor, resigned. He was concerned about the safety procedures.

Jose Garcia, a 30-year veteran at the Kennedy Launch Center in Florida, wrote to President Clinton in August 1995, expressing doubts about the upheavals within NASA since Goldin's arrival. He posted his letter on the Internet, challenging other staffers with doubts to go public: 'Managers who have concerns are leaving the program... It's disappointing that some of them don't have the fortitude to speak their mind.'

A close colleague of Goldin's has said, 'Dan is a really superb manager of technology, but he doesn't count the human costs. He doesn't realize people's careers are being destroyed. If you have a big organization like ours, and a lot of people are unhappy, then that's not a good sign.'

These are tough times, as the space agency struggles to find a new identity. 'I know it's hard,' Goldin explained recently, 'but you tell me how to maintain confidence among your people when you have to absorb budget cuts. We should be a $22 billion agency, and we're going to have to operate on $10 billion. It's not easy.'

And this is true. It is not Goldin, so much as President Clinton and the government that have insisted on a leaner, cheaper NASA. Currently the agency is spending a total of $14 billion annually. Clinton wants that reduced by 30 per cent before the year 2000, space station, shuttle and all. His support for NASA is dependent on such a commitment.

**A FORTHRIGHT MAN**
Daniel S. Goldin has brought a very different management style to NASA. This photo portrays him at his confirmation hearing before the Senate Committee on Commerce, Science and Transportation (March 27, 1992) during which he was formally appointed to lead America's space effort.

**PRIVATE ENTERPRISE**
The American government is keen for private industry to absorb more of the costs of space activity. The Spacehab company was formed in 1983 to create a commercial module for the space shuttle. The idea was that other private buyers should rent flights aboard the module for microgravity research. Spacehab raised $100 million, the largest private sum ever invested in space outside the satellite industry. Customers are offered space for experiments in compact lockers, as shown in this artwork. The first Spacehab flew in June 1993, but other companies were more cautious than the Spacehab shareholders had hoped. NASA have become the major client. The basic problem is hard to resolve: where is the profit for private companies in space, unless government agencies are prepared to hire them?

## REDESIGNING THE REDESIGN

William 'Bill' Clinton came to the White House in 1993, the first Democrat president for 12 years. The House of Congress, however, continued to be dominated by Republicans. In particular, 'Newt' Gingrich, the House Speaker, established himself as a powerful figure. Throughout his first term Clinton was caught up in a tricky balancing act, on the one hand trying to live up to his Democrat ideals, and on the other, appeasing the Republican anti-spending lobby in Congress. During

January 1996 the government actually closed down for two weeks, because Congress refused to release funds for federal workers' salaries unless Clinton agreed on cutbacks to allay the huge national deficit. Clinton stood his ground, Gingrich lost some, but the arguments continued. And in the midst of all this, NASA continued to plan the multi-billion-dollar space station. With social and medical federal programs already under great strain, NASA were a soft target in the budget battles.

When he first entered the White House, Clinton did not see space as a priority. He had not been expected to retain Dan Goldin, a 'leftover' from former president George Bush's administration, but Goldin had been in harness for less than a year, and nobody could justify removing him before he'd had a chance to make his mark—for better or worse. Anyway, he was promoting his new message about space: 'faster, better, cheaper'. That was his trademark credo, and his reorganization efforts found favour with Clinton, who asked him to stay on as NASA's chief. Goldin's skills as a space salesman also helped him convince even his harshest critics in Congress to give the space agency another chance.

NASA's big, fat space station project did not sit so well with the politicians, however. On March 9, 1993, barely into his presidency, Clinton declared an ultimatum. The station must be totally redesigned within 90 days, to slash its costs. Congress had made a similar 90-day demand back in October 1990, but Clinton and his advisors were more specific this time. They wanted to see three options, costing $5 billion, $7 billion and $9 billion respectively. Even the highest figure on offer amounted to less than half the cost of NASA's existing proposal.

An advisory panel, the 'Vest Committee' (named after its Chairman, Dr Charles Vest) was established a fortnight later, to assess NASA's proposals. Meanwhile, Goldin had taken 45 of his best station designers and set them to work, along with representatives from the international partners—a wise move, but not enough to ease their anxieties about the Vest Committee's likely impact on their plans.

On June 10, the Vest Committee reported its findings to President Clinton. NASA had indeed presented three new options, as requested. Two of them, option 'A' and option 'B' consisted of trimmed-down variations of the exisiting Freedom design.

A third concept, option 'C', was compact, relatively cheap and much simpler. It threatened to wipe out most of Freedom altogether, in favour of something completely different. This would have been a huge shock to all the management, commercial and diplomatic interests that had grown up around the project.



## OPTION 'C'

Option 'C' consisted of a large module, stored inside a canister slightly fatter, and rounder in cross-section, than a shuttle orbiter, with shuttle rear engines, shuttle solid rocket boosters and a standard drop-away fuel tank. But this 'Shuttle-C' variant would have no wings, no landing gear and no crew cabin. Instead of just packing station components into the 4.6m-wide (15ft) cargo bay of an exisiting orbiter, every last scrap of available weight and interior space would be devoted to the payload. It would mean creating a new booster out of shuttle components, but the engines, fuel tanks and so forth were already available, and the costs of conversion could be offset by launching most of the space station in a single flight.

NASA's shuttle managers were concerned about the implications for their fleet of four winged orbiters. *Columbia*'s rear motor assembly and tail section would have to be sacrificed for inclusion in the new unmanned booster, while the wings and crew cabin languished in storage. The manned fleet would be reduced to three ships; but in comparison to the Freedom-derived space station concepts, option 'C' would have required fewer supporting shuttle flights overall, so *Columbia*'s sacrifice might still have been worthwhile. In many ways 'C' was the best possible design; cost-effective, relatively free of risks, expandable in the future, and requiring very few dangerous spacewalks to bring the station on-line: a safety hazard that had featured as a black mark against all the other designs to date. 'C' had the added bonus of stimulating development of a new unmanned booster, with a heavier lift capability, from existing components.

Most of the contractor companies were opposed to option 'C'. The pressurized laboratory modules and living compartments planned for Freedom would be reduced to one 'big can' and a couple of add-ons. The sprawling truss elements of the current design would be eliminated. NASA's foreign partners were also alarmed. Option 'C' would require a total rethink of all the work they had done so far. The Vest Committee found that option 'C''s 28m-long (92ft) core module 'provides more volume than the current Freedom baseline, thus putting into question the real need for additional volume provided by the International Partners'.

## PASSING ON 'C'

The eventual decision not to adopt option 'C' demonstrates very clearly how the station project was only partially concerned with its ultimate purpose—whatever that was supposed to be. After all this time, its usefulness still had not been properly defined. The hardware design was being driven largely by the needs of the contractors and space agencies, and rather less by any valid mission requirements. But 'C' was the only design that even came near the White House's cost targets, with its price tag of $15 billion. The Committee warned Clinton, 'The costs can only be minimized if [your administration] makes a firm commitment to provide stable funding … None of the three options meets [your] costs targets while simultaneously achieving the milestones desired … [Your proposal for] funding fails to recognize the inherent staffing profiles associated with such programs.'

Telling a president that his cost recommendations were at least in part 'wrong' was quite a bold thing to do, but the Vest Committee would have failed in their job if they had pretended a space station could be built for $9 billion or less. Their minimum figures ranged from $15 billion for option 'C' to $19 billion for a reworked variation of Freedom. Just as

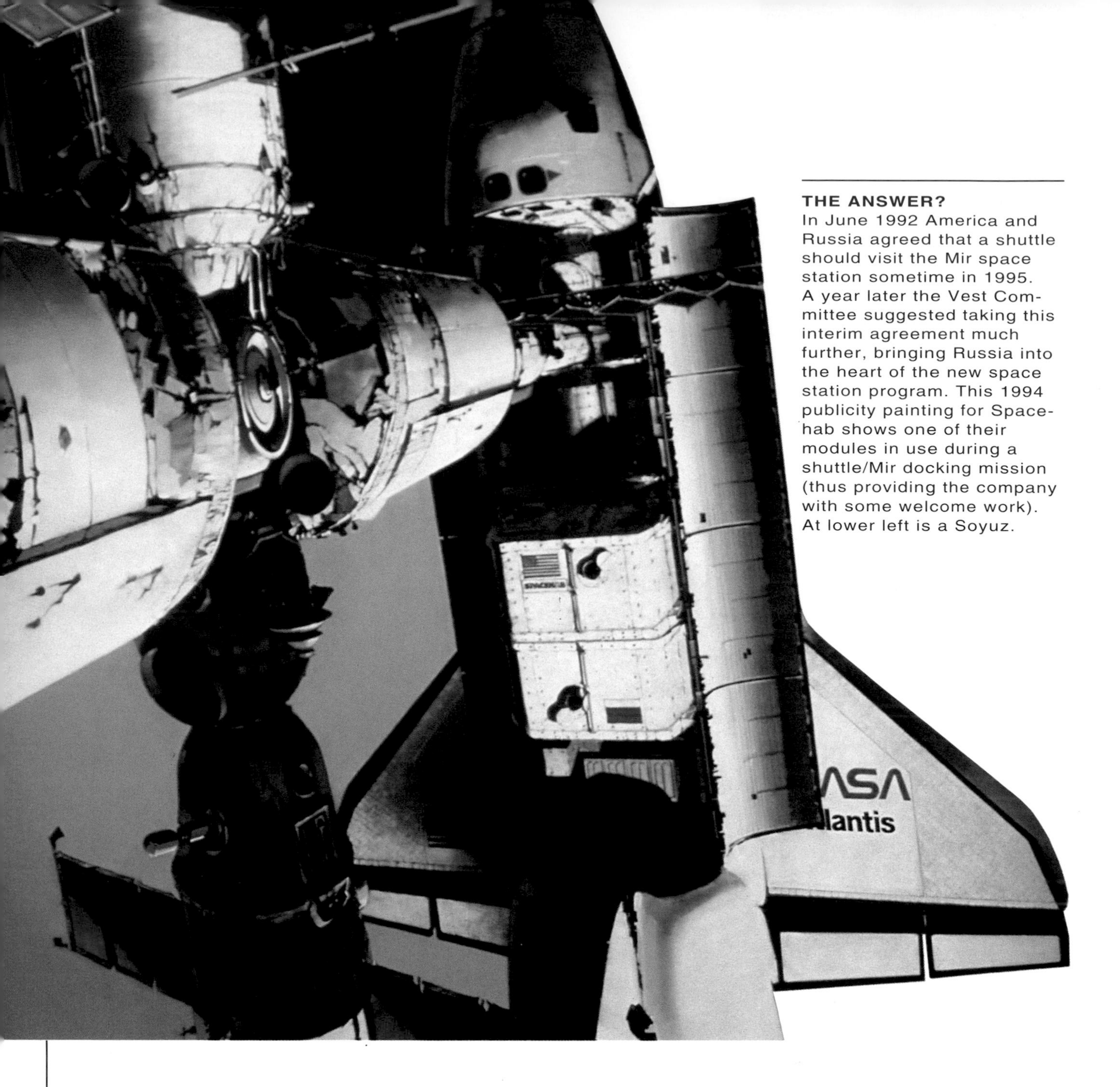


**THE ANSWER?**
In June 1992 America and Russia agreed that a shuttle should visit the Mir space station sometime in 1995. A year later the Vest Committee suggested taking this interim agreement much further, bringing Russia into the heart of the new space station program. This 1994 publicity painting for Spacehab shows one of their modules in use during a shuttle/Mir docking mission (thus providing the company with some welcome work). At lower left is a Soyuz.

Weinberger had told Nixon in 1971 that NASA could not simply be ignored, so the Vest Committee politely challenged Clinton to make up his mind. Did he or did he not want America to build *some* kind of a space station? If so, he would have to support a realistic funding package.

Two weeks after the Vest Report, the White House decided to opt for an amalgam of option 'A' and option 'B', essentially yet another version of Freedom. President Clinton asked Congress to approve NASA's upcoming budget requests. The lobbying was intense. And it had to be. On June 23, the space station's funds for 1994 were approved by 215–216. Just a single vote saved the program from death. After the poll, Dan Goldin appeared on the steps leading up to Capitol Hill, and spoke to waiting reporters. His smile seemed rather strained. 'A win is a win,' he said.

Behind him, Republican representative Dick Zimmer and Democrat Tim Roemer gave their version of events. They had been the station's most persistent opponents, and had led most of the efforts to cancel it. Roemer said, 'We'll kill it. If not this year, then next year, but it'll be killed in the end.'

Meanwhile, a rival technology project, a 'Superconducting Supercollider' in Dallas, Texas, had been cancelled by Congress only the previous day, by a massive margin of 130 votes. The Supercollider had been intended to push knowledge of subatomic physics to new levels. The science community was appalled that its 'pure' research should have lost out to a space project with less obvious scientific value. But congressional mathematics was based on different criteria than the merits of subatomic discovery. NASA's space station had already absorbed $11 billion since 1984, along with $3 billion of foreign commitments. In contrast, the Supercollider had spent only $1.7 billion, with no foreign involvement. Congress did not wish to be accused of squandering the money that NASA had already spent. Maybe the Dallas scientists should have wasted more cash ahead of the Supercollider vote.

## OLD FOES, NEW FRIENDS?

So, the space station had survived, but this was as far as it could afford to fall. For Congress, the Vest Committee material seemed like one redesign too many. Obviously NASA needed to pull a new rabbit out of the hat, if ever their station was to make some real progress. And by now, there was indeed another rabbit in the wings. The Vest Committee recommended bringing in a new partner. 'All three design options presented here are dependent upon the space shuttle. This is undesirable from the perspective of risk, and should be ameliorated by using an alternative expendable launch vehicle, such as the Russian Proton… All options have a firm requirement for a "lifeboat"… The Russian Soyuz can easily be attached... The Committee feels that Russia has important capabilities that could be advantageous…'

Here was yet another new shock for ESA, Japan and Canada to absorb. Who, exactly, would benefit from Russian participation? Would their involvement generate another redesign? Would their ailing economy put the whole project into jeopardy? Dan Goldin obviously didn't think so.

On June 6, 1992, Dan Goldin and his Russian opposite number, Uri Koptev, met for the first time in Goldin's Washington apartment. Their meeting was not exactly secret, but neither was it reported in the press. Such caution was understandable. At this early stage, neither man knew what he would find in the other. Despite barriers of nationality, history and language, Koptev and Goldin quickly saw how similar their problems were. They were both ex-Cold Warriors who believed in exploring space for peaceful ends. They had both been stripped of much power by their political masters, and forced into painful spending cuts. Dr John Logsdon, a space policy advisor in Washington, describes this meeting as 'like the two of them falling into "administrator's love". They were almost mirror images of each other.'

**GOLDIN'S RUSSIAN TWIN**
Yuri Koptev, General Director of the Russian Space Agency.

## THE BROKEN-BACKED BEAR

Goldin and Koptev's meeting was not entirely without diplomatic precedent. Since the collapse of the old Soviet Union, Russian and American presidents have consistently discussed joint space activities at their regular Summit Meetings. Ronald Reagan made a private approach to Russia in advance of his 1984 State of the Union speech, inviting them to participate in a simulated rescue mission between a shuttle and Salyut 7, which was still in orbit that year. Reagan's offer fell far short of an invitation to work on the new space station itself, and at that time, Russia did not feel their Salyut crews would ever need rescuing by NASA, so they were cool to the offer.

By June 1992, the situation was very different. Post-Soviet Russia was now thoroughly preoccupied with getting rescued, not so much through diplomatic dockings, but with dollars. Within days of Goldin and Koptev's private meeting at the Watergate, Presidents George Bush and Boris Yeltsin agreed to make plans for sending an American shuttle up to Russia's Mir space station, and also to fly Russian cosmonauts on a shuttle. This was a real diplomatic advance, yielding the first hope for joint space flights since the Apollo–Soyuz docking 17 years earlier.

While the Soviet Union still existed, most space activities fell under the jurisdiction of the elegantly named Ministry of General Machine Building. There was no NASA-style independence, and the notion of a 'civilian' agency was out of the question. At its height, the space program employed 600,000 people, many of them wearing uniforms. The manned Soyuz vehicles, the Salyut stations and the current Mir had contributed to an almost non-stop presence in space, and by 1991, Soviet cosmonauts had clocked up 23 years' total flight time between them, while their American rivals had accumulated just 8 years. Despite escalating economic problems, Soviet launches, including all the unmanned probes and satellites, Progress ferries and so forth, averaged one every five days.

The decay set in very rapidly. In May 1991, the Mir space station had been operational for five years and was starting to creak at the seams. Flight engineer Sergei Krikalev was sent up on a six-month tour of duty. He conducted a number of space walks and worked through a busy schedule of repairs. As October approached, Krikalev looked forward to the docking of another Soyuz bearing a relief engineer, so he could go home to rejoin his wife and baby daughter. But some of the messages from ground control were making Krikalev uneasy. Tactfully, they

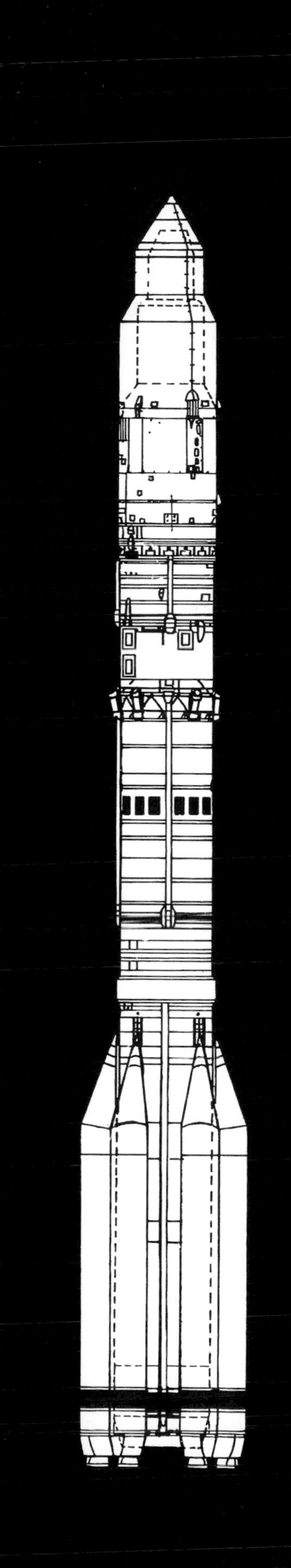

**BAIKONUR'S BOOSTERS**

At left is the 50m (164ft) Soyuz SL-4 rocket, consisting of two central stages and four 'strap-on' liquid-fuelled boosters. The Soyuz capsule rides in the top shroud, which is capped by an escape motor. The SL-4 first flew in 1963, and it remains the primary launch vehicle for all Russian cosmonauts.

At right is the unmanned heavy-lift booster Proton SL-13, a 55m (180ft) vehicle that carried Salyut and Mir components into orbit, and now serves as an effective commercial satellite launcher. A Mir core module can be seen in outline within the payload shroud. Proton is assisted by six solid boosters.

explained that a special guest astronaut was scheduled to visit Mir in November, and there wasn't enough money for that flight. So they would bring up the guest on the October relief flight, instead of the properly trained engineer they had promised. Krikalev was asked to stay up another five months.

Krikalev was used to guest astronauts. On his flight up to Mir he had sat next to British astronaut Helen Sharman, who had stayed on the station for a few days, subsidized by a London-based consortium. More than a dozen foreigners had visited so far, including a Japanese journalist who spent much of his stay nauseously gripping the walls, and serious-minded European astronauts who got on with genuine science programs. Ticket prices for these visitors ranged from $8 to $12 million, depending on their length of stay. Recently, it had become almost impossible to schedule any Soyuz launches unless a foreigner was paying for the flight.

But this latest guest was not quite as foreign as Krikalev might have expected, and neither would he be paying a fare for the ride. It seemed the space authorities in Russia had lost control over their remote launch site at Baikonur, in Kazakhstan, because that once-tamed dominion in the south of the Soviet empire had shrugged off Moscow's dominion and declared itself an independent country. Kazakh authorities were now demanding that Russia make the goodwill gesture of sending a Kazakh citizen for a visit to Mir at the first opportunity, or else Russian space workers would be denied further use of the Baikonur complex. And there was more unsettling news. The mission controllers weren't too sure how to tell Krikalev about the tanks that had rumbled through the streets of Moscow that August, during a coup attempt that ultimately unseated the liberal reformer Mikhail Gorbachev but failed to reinstate the old regime as had been intended. In fact the coup had the opposite effect, destroying the last vestiges of communist rule and shattering the Soviet empire. Krikalev's mission badge proclaimed him a cosmonaut of the Union of Soviet Socialist Republics. As from now, that place or state of mind no longer had any sensible meaning.

Neither did mission control highlight the fact that two of the coup's leaders, Oleg Baklanov and Vitaly Doguzhiyev, had been senior figures in the rocket program. With this in mind, Russia's new leader, Boris Yeltsin, had become ill-disposed to the space people. In February 1992 he abolished Krikalev's employer, the once-great Ministry of General

Machine Building, replacing it with smaller and less powerful entities. Rocket activities were now the responsibility of a new specialized authority, the Russian Space Agency (RSA). Commercial activities, the hunt for foreign hard currency, now had precedence over the old military posturing. Immediately RSA was plunged into the same kind of struggle that NASA had faced: to keep the army generals from hijacking their dwindling financial resources.

## STATES OF CHAOS

When Krikalev finally came home in March 1992, he had spent 313 days on a tiny, unchanging communist-built island in space, while his vast homeland had torn itself to pieces and changed forever. He found to his dismay that the back salary for his long and tiring mission was hardly an adequate reward for his services, equating in near-worthless roubles to maybe a couple of hundred American dollars for ten months' work. He and his family could barely afford food. Krikalev's new employers, the RSA, were in no position to help, since their finances had collapsed before they could even get started. None of them could afford food. A colleague of Krikalev's summed up the situation to a Western journalist: 'Once, the government would throw more money at us than we could eat. Now we wonder about our next meal.'

**STURDY BOOSTER**
A typical Soyuz launcher and its capsule in a protective white shroud being moved into its firing position at the Baikonur Cosmodrome during February 1992. A few hours later it is upright against the gantry. The white-painted booster in flight (*far right*) carries Norman Thagard, the first American visitor to Mir, on March 14, 1995. The reliability of the Soyuz rocket is a positive factor for a space program that often seems on the verge of collapse.

The jubilant Kazakh authorities had gained a partial stranglehold over a rapidly diminishing asset: Baikonur itself. The nearby town of Leninsk had grown up around the center over the last 40 years, and now, tens of thousands of its citizens, mainly skilled Russian and Ukrainian technical staff, were packing up their possessions and moving away like war-torn refugees, because Baikonur no longer had jobs for them. The remaining Kazakhs had always been treated as second-class citizens in their own land. Russia had never trained them in the engineering skills required to launch rockets—or even to keep Leninsk's power station going. Resentments boiled over. Discarded first-stage boosters from four decades of Russian launch activities were scattered across vast areas of Kazakh farmland, leaking thousands of gallons of unspent fuel into the ground. Under Soviet rule, nobody had dared complain. Now the locals demanded that Russia should clear up this toxic rubbish and surrender half its ownership of Baikonur. Meanwhile, the newly instated Kazakh border customs hijacked Russian machinery on its way to the launch center, and held much of it for ransom.

In February 1992, just as the newly founded RSA was trying to forge some kind of working relationship with the Kazakhs, Baikonur's army guardians rioted in the first of several rebellions stemming from lack of food, poor salaries and the squalor of their living conditions. They burned their barracks to the ground, and three people died in the chaos.

**LIFE ON MIR IS NO JOKE**
The Mir's core module, launched in April 1986, is starting to show its age, rather like the rest of the Russian industrial and economic base. NASA shuttle crews like to joke that cosmonauts clamour to fly up to Mir so as to improve their average standard of living. This seems unlikely.

A year later, the RSA's problems at Baikonur had become catastrophic. In May 1993, a rocket bearing a communications satellite fell into the Pacific like a damp squib, because there hadn't been enough clean fuel available for Baikonur's gantry crews to fill the booster's tanks. A month later, 500 Kazakh conscripts stormed the center, smashing equipment, hijacking Russian workers' cars and causing massive damage.

## DISUNITED STATES

As if the Kazakhs were not awkward enough, certain key rocket and space facilities in the Ukraine played their independence from Russian hegemony for all it was worth. In addition to providing booster components, the Ukraine hosted several communications relays that kept Moscow's mission controllers in touch with orbiting cosmonauts. Sometimes the Ukrainians would blackmail Moscow by switching off their radio dishes when Mir was flying over their territory. On one occasion, they cut the link during a spacewalk.

At least the Ukraine had some economic and political bargaining power. Kazakhstan had nothing but Baikonur itself. On March 28, 1994, with the prospect of foreign commercial interest in Russia's remaining space capabilities forcing him into action, President Yeltsin signed an agreement effectively renting Baikonur for the equivalent of $110 million a year, paid in the form of trade credits. A few weeks later, a fire at the center destroyed rocket components in an assembly shed. There was no water available to fight the flames.

It is hardly fair to compare subjugated nations like Kazakhstan with any Western equivalent, but imagine a parallel: suppose California, Texas, Florida and all the other American states suddenly declare independence, while an ultra-Republican coup puts tanks onto the streets of Washington. Then expand the usual in-fighting between NASA and the military, and throw in the many rival contractor companies, to produce a mélange of bickering and confusion. Imagine NASA broken up to neutralize their political influence, and the various field centers now competing for control over what little space activity remains. Imagine persistent rioting and looting at the Kennedy launch center, while Floridian authorities impose heavy taxes on hardware arriving from other states to be assembled for flight. Imagine that Florida then demands that the Johnson control center in Texas pays a massive levy to use the launch center at all. Add to this the poisoned spice of near-endemic bankruptcy, causing thousands of job losses, and plunging the value of space workers' salaries from $50,000 a year to maybe a couple of thousand dollars at best, and the running of an effective agency becomes impossible.

Now imagine how low your space workers' morale would sink under such circumstances. A rocket ferry bound for an operational space station might, for instance, be looted on the launchpad, stripped of its food packages by ground crews desperate to feed themselves, or to sell the ferry's contents on the black market for hard currency. Imagine the ferry blasting into the sky half emptied of its supplies ... In mid-1994 a Progress capsule going up to Mir was indeed pillaged just before launch.

The chaos at Baikonur was not the only problem facing the RSA. Within Russia itself, surviving factions of the old Ministry of General Machine Building squabbled over who was in charge of space. RSA was responsible for mission control, and all the flight activities. Meanwhile, NPO Energia, a factory complex, retained control over the construction of Soyuz capsules. It didn't help that many of their assembly facilities were housed at Baikonur, deep in rebellious Kazakhstan. Meanwhile, another powerful manufacturing facility, Khrunichev, built all the Proton boosters for launching Mir hardware.

In February 1992, Yuri Koptev was given the unenviable task of trying to sort out this mess. He had been second in command at the old Ministry of Machine Building, but was untainted by any association with the failed coup. President Yeltsin selected him to lead the new RSA. This big, burly man with a gruff manner and a steely gaze was perhaps the most important unifying factor as the space program struggled to reassemble itself into a sensible whole. Koptev encouraged American aerospace firms like Lockheed and Boeing to set up offices in Russia and strike alliances with local factories. These large foreign corporations proved to be valuable allies. They came armed with hard currency instead of useless roubles. Their bargaining power enabled them to impose market disciplines on the Russians. But it was a strange trading environment, which at first left the Americans confused. In the Soviet days, factories never worried about profit-and-loss, or even about financial control. They just fudged their figures to fit in with the State's latest five-year plan. Now, with foreigners demanding to know what they might charge for certain goods and services, Russian managers had no idea how to price their products. Western negotiators also found that the notion of a 'contract' was alien to Russian business managers, who had seldom come across them in their dealings with the old State.

Yuri Koptev turned increasingly to the West in search of a solution to his problems. He was very surprised and gratified to find that NASA's Dan Goldin seemed to need him almost as much as he needed Goldin. NASA needed to find extra impetus for building their space station, and Russia seemed capable, at least by default, of supplying it. The station would now become a significant tool of East–West relations—a stabilizing factor between the old superpowers, and a means of easing tensions.

**CRAMPED CAPSULE**
Two cosmonauts in a Soyuz simulator at Star City, the major Russian space training center near Moscow. Soyuz can carry a crew of three, but one of the seats can be replaced by a small cargo container. Crews must wear pressure suits during launch, making the cabin an even tighter fit than in this photo. Soyuz is a throwback to the 1960s when all space travellers had to squeeze into tiny capsules.

**WASHINGTON, SEPTEMBER 1993: THE DEAL BETWEEN RUSSIA AND AMERICA THAT ENDED THEIR RIVALRY IN SPACE**

**Phase One:** *The Shuttle-to-Mir mission first agreed by Presidents George Bush and Boris Yeltsin on June 17, 1992, and scheduled for 1995, is extended to include further shuttle dockings up to 1997, with American astronauts scheduled for a total of two years' flight time aboard Mir. Shuttles will help service and resupply the Russian station. Two additional Mir modules, 'Priroda' and 'Spektr', will be equipped for American scientific experiments.*

*For Phase One activities, Russia receives $100 million during 1994, and a further $100 million per year through 1997.*

**Phase Two:** *Russia will adapt some of their existing unlaunched hardware as a building block for the International Space Station. This 'Functional Energy Block' (FGB) will provide interim propulsion and power while the rest of the station is prepared for assembly around it. Russia's unmanned Proton booster will be made available free of charge for launching the block and other components.*

*In return for these elements, Russian industry will be compensated at an appropriate rate to be agreed in the near future.*

*Russia and America will jointly initiate design of a common space suit, life support and environmental control systems. A solar furnace power system will be jointly designed and tested on Mir.*

**THE OTHER SIDE OF THE BARGAIN: SECURITY OR PROTECTIONISM?**

In a supposedly separate set of contracts, American negotiators extract a high price in return for Russia's $100 million a year:

*Russia is permitted to launch up to eight commercial satellites for non-Russian clients, up to the year 2000, provided that launch fees are compatible within a 7.5 per cent margin with European and American prices for similar launch services.*

*Under the terms of the Missile Technology Control Regime (MTCR), Russia will abandon current plans to sell liquid-fuelled rocket technology to India.*

This agreement put Russia in a strange position, effectively allowing their European and American competitors to determine what was a fair price for satellite launches, thus denying Russia the privilege of undercutting rivals without limit.

## SIGNING THE DEAL

Now that President Clinton was in the White House, the pace of Russian–American integration accelerated further. When Goldin convened his 45-strong team in March 1993, to redesign the space station ahead of the Vest Committee, he brought in some Russian engineers to observe the proceedings. With little formal preamble, they sat down with their American counterparts to talk nuts and bolts. What if Russia helped build a space station? How might it be done? Just as Europe's Spacelab engineers had cut through the flab of politics to form a rapport with American engineers, so the Russians made friends with NASA people who spoke the same language of orbits, trajectories, power-to-weight ratios, docking collars, engine thrusts and cooling systems.

The positive results of these and other sessions fed their way into the Vest Committee's recommendations of June 1993 to bring the Russians on board. On September 2, Vice-President Al Gore, Russian Prime Minister Victor Chernomyrdin, and the two space chiefs, Koptev and Goldin, signed a major deal at a special meeting in the White House. Russia would become involved in the space station. From now on, in deference to Russian sensibilities, the project would no longer be called 'Freedom', but 'Alpha'. The press heralded the event as a historic pact. Some of the reports took the trouble to mention a secondary set of agreements signed that day, which banned Russia from selling space technology to India and limited their ability to sell cheap satellite launches in direct competition with American and European commercial operations.

The motives behind all these agreements were complex. Just as James Beggs had used the prospect of European collaboration in the 1980s to protect NASA's space station in Congress, so Goldin now wanted to tie the station's fortunes into the geopolitical relationship between America and Russia. Goldin has consistently claimed that Russian involvement will speed up the station program and save NASA money. These claims were by no means dishonest, but neither were they verifiable. Congress recognized that the true value of Russian involvement lay elsewhere. In particular, the American political establishment saw collaboration in space as a way of helping to stabilize their relations with Russia as a whole.

Although the Cold War had ended, Russia remained unpredictable, and still heavily armed. Their economy was in a shambles, and the temptation to sell weapons technology abroad was overwhelming. With Goldin and Koptev having formed such a close rapport, an opportunity arose for America to gain a measure of control over Russia's unstable technology base. Helping to keep Russian engineers employed on peaceful international space projects would lessen the temptation for them to sell their expertise to Libya, Iraq and other undesirable clients, or so the theory went. Hence, the intimate link between the space station agreement and the limiting of Russian sales of rocket technology to India.

Indian space officials were rather perplexed. They never wanted Russian rocket engines for missiles. They just wanted to launch satellites. In part, American anxieties stemmed not just from the potential spread of weapon technology, so much as the threat of competition in the commercial satellite launch market.

But if the Washington diplomats thought they had pulled off a masterstroke with these deals, other people were not so sure. The *Moscow News* wrote: 'We are selling out our space technologies, providing breakthroughs to the Americans almost for free.' And in the West, the highly respected science journal *Nature* warned: 'The biggest question mark hangs over the wisdom of allowing the fate of an international program that has already been almost destroyed by the vagaries of the US Congress to depend on the even greater vagaries of Russian politics.'

Goldin and his team of negotiators visited Moscow during the first week of November. As they sat down to talk with their Russian hosts, the windows of their conference room started to shake. Again the tanks were on the streets of Moscow, and they were firing in earnest. President Yeltsin was storming the rebellious Parliament Building.

**A SPRAWLING SPACE COMPLEX**

This drawing depicts a docking between an American shuttle and the Russian Mir station. Protected by weightlessness, the delicate structures point in all directions. The weight and launch date of each Mir module is as follows:

**Core Module** (February 1986) 20,400kg (44,880lbs), contains the main cabin, and includes a Soyuz docking collar at the rear, with a multiple port for additional modules at the front.

**Kvant 1** *Quantum* (March 1987) 11,000kg (24,200lbs), provides additional interior workspace.

**Kvant 2** (November 1989) 19,565kg (43,043lbs), contains a urine-regeneration system, a shower and an airlock to support space walk activities.

**Kristal** *Crystal* (May 1990) 19,640kg (43,208lbs), carries additional docking collars and science payloads, including materials-processing furnaces.

**Spektr** *Spectrum* (May 1995) loaded with 755kg (1661lbs) of experiments for American guest astronauts.

**Priroda** *Nature* (November 1995) was added in April 1996, carrying 935kg (2057lbs) of American experiments. It is not shown in this diagram, which shows the second shuttle/Mir docking of November 1995, prior to the Priroda's arrival.

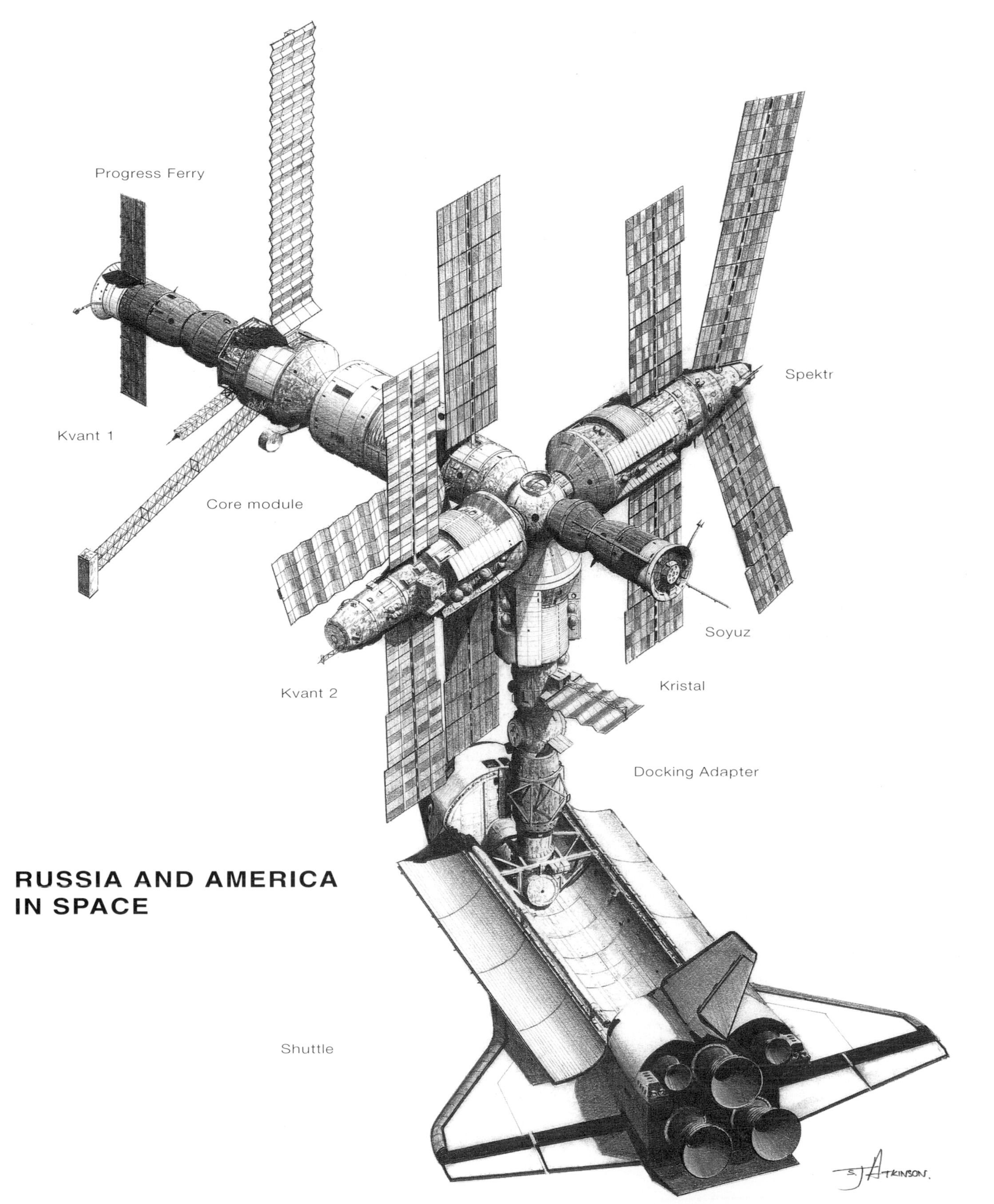


## RUSSIA AND AMERICA IN SPACE

# THE ALPHA BET

*Between 1993 and the end of 1995, the space station project, now called 'Alpha', gained technical and political momentum. NASA gained some valuable successes in its shuttle operations, and the new alliance with Russia, although apparently fragile, began to deliver solid results in space. The intervention of an Oscar-winning actor also proved a bonus for Alpha's fortunes.*

**IN ORBIT AT LAST?**
NASA's computer graphic simulation shows the International Space Station in its complete form. The first elements will be launched on a Russian booster in November 1997, and the station should be complete by the year 2002.

NASDA
USA
Atlantis

# THE ALPHA BET

Just before Christmas 1993, something happened that put the space station back on track. NASA did something spectacular. Something that worked. Something that looked good on television, pleased the politicians, and diverted attention from the international tensions.

The Hubble Space Telescope was still in a mess. Since its launch in April 1990 the problems had piled up. The mirror was the wrong shape, and the huge solar panels that powered the on-board equipment flexed and buckled every time Hubble's orbit took it into, or out of, direct exposure to the sun. Since these and other problems had become evident, NASA had been working out how to fix all these billion-dollar bungles.

Fortunately, Hubble was designed right from the start for in-orbit servicing by shuttle astronauts. Optical engineers realized that the faulty mirror could be corrected at the 'eyepiece' end of the assembly, using a clever system of lenses to bend the fuzzy images back into focus. Meanwhile, the British company who had manufactured the solar panels worked out why they were behaving so poorly, and built a new pair of panels that would remain stable during temperature fluctuations.

On December 2, 1993, shuttle *Endeavour* lifted off, carrying a rescue crew for Hubble. Working as two separate duty teams, astronauts Story Musgrave, Jeffrey Hoffman, Thomas Akers and Kathryn Thornton completed the repairs in five back-to-back spacewalks totalling 35 hours. Hubble was pulled into *Endeavour*'s cargo bay using the Canadian Remote Manipulator robot arm (a crucial piece of equipment that has featured successfully in many shuttle flights). Musgrave and Hoffman manhandled Hubble into a secure position and began replacing equipment. Then Thornton performed a dangerous but necessary manoeuvre, using a pair of shears to snip off the old solar panels, while *Endeavour* sent out a brief burst of thruster fire to send the panels safely down into the atmosphere.

After this, Thornton and Akers loaded a self-contained package of corrective optics into a side hatch on Hubble. During subsequent spacewalks, the crew added more instruments and new gyroscope control systems, before gently easing Hubble free of the *Endeavour* and back into an independent orbit. Hubble's instruments may have been flawed, but the main spacecraft structure turned out to have been well designed, for once. The access doors opened properly, and the new equipment slid into

LONG-DISTANCE VOYAGER
Nearing the end of his record-breaking 438-day mission, cosmonaut Valeri Poliakov looks out of Mir's window in February 1995 to watch the shuttle *Discovery*'s approach, and the shuttle crew looks up at him. (The shuttle did not dock on this occasion.)

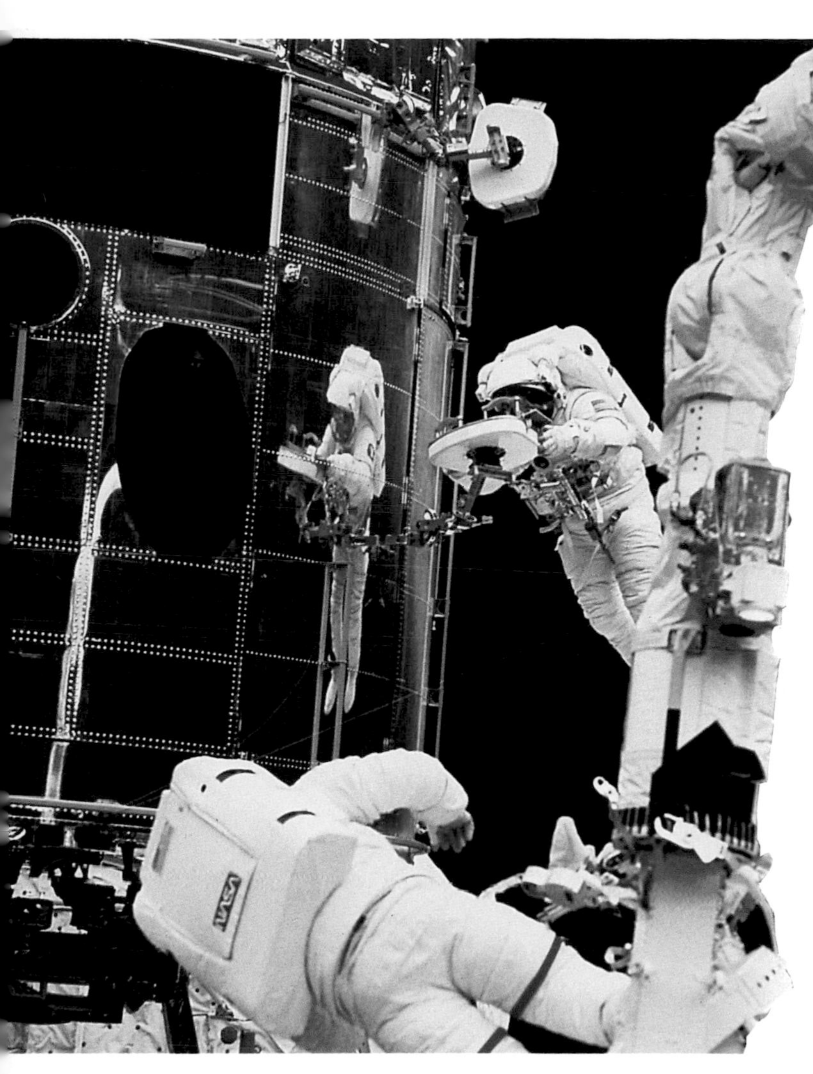


place according to plan. The repair mission epitomized what always used to be known as NASA's 'can-do' skills. The spacewalks were spectacular, and NASA's new-found confidence seemed infectious. Some of the agency's congressional critics began to revise their views, perhaps unwilling to appear ungracious in the light of an obvious success. The complex spacewalks also demonstrated that building a space station need not be as foolhardy a scheme as some commentators had imagined. Alpha's planners were now armed with a conclusive precedent for bolting components together during space walks.

Then, in February 1994, the redoubtable Sergei Krikalev became the first cosmonaut ever to fly aboard an American shuttle, *Discovery*. Russian–American collaboration began to feel like less of an abstraction. In Congress, the space station faced its by-now routine challenge. Representatives Dick Zimmer and Tim Roemer were on hand together with their usual bipartisan alliance, backing a motion to kill it. The vote came up on June 29. But this time, the mood was different. A year before, the station funding had survived by just one vote. Now the safety margin was 123.

Goldin felt it was time to move beyond simply getting the year-by-year budgets passed. Instead of merely surviving motions-to-kill, the station should seek positive approval—a formal long-term commitment, just as the Vest Committee had recommended. Goldin really needed this assurance, because six months earlier he had signed a definitive deal with a single major contractor to ramrod the space station towards completion. That deal was worth $5.6 billion, and at this late stage its beneficiary, Boeing, could not afford any nasty surprises in Congress.

**MISSION ACCOMPLISHED**
Astronauts repairing the Hubble Space Telescope in December 1993, during the shuttle *Endeavour*'s fifth flight. Kathy Thornton's face can be seen inside her helmet (*left*).

## BOEING IS BOSS

On January 13, 1995, Boeing signed a $5.6 billion contract with NASA to become prime contractor for Alpha, thus centralizing control over the various companies competing for slices of the station cake. From now on, the endless squabbling would cease, and Boeing would call the shots. Rockwell, McDonnell Douglas, Lockheed and other companies who had hoped for a bigger role were comforted by substantial subcontracts. Dan Goldin appointed NASA's Johnson Space Center in Texas as the unchallenged leader of the project, and eliminated all the duplicated activities in other centers.

A month after signing the NASA deal, Boeing firmed up a $190-million contract with Russia's Khrunichev manufacturing facility for purchase of a special propulsion unit, a 'Functional Cargo Block' (usually given its Russian acronym, FGB). That word 'functional' is crucial. In the early stages of space station construction, FGB will provide the thruster power required to stabilize its orbit, along with a basic electrical capability. Russia may be building most of the hardware for FGB, but Boeing has accepted ultimate responsibility for delivering it on time.

The consolidation of their position made Boeing's bosses very happy. In part, NASA's decision to give them the prime contract may have been influenced by innovative management practices instituted during design and production of their latest wide-body airliner, the 777. But there was something else which the company had in their favour: a ready-made facility deep in the heart of a NASA field center. During the Apollo days, Boeing had constructed the gigantic first stages for the Saturn moon boosters. They built a substantial management and assembly facility in the heart of NASA's Marshall Center at Huntsville. In 1966, they employed 4,500 staff there. (The largest Saturn components were built at Michoud, from where they could be taken down to the Kennedy Center by barge, but these activities were administered mainly from Huntsville.)

By the mid-1970s, this staffing level plummeted to perhaps a couple of dozen people. Boeing was forced to lay off most of its Apollo workers and divert financial resources into other plants to support production of their 747 Jumbo Jet airliners; but they made a bold decision to hang on to the huge, expensive, empty facility at Huntsville, waiting for the day when their fortunes in the space industry might change for the better.

Today, more than 1,000 staff work in the plant, building space station modules. The work has gone well—in fact, the modules are very nearly completed. They just have to be tested, fitted with interior equipment by the McDonnell subcontractors, and then sent off to Florida as payloads for gradual integration into the shuttle launch manifest. So what will happen to Boeing's workforce in a few years' time? No doubt their Space Group will be lobbying in Washington any day now for something new to build.

**ASSEMBLY COMPLETE**
November 1995: The American Laboratory Module on completion of welding at Boeing's Hunstville facility. The waffle texture adds strength and rigidity. Thermal and micrometeoroid blankets will be placed over the waffle grid prior to launch. The module is 8.5m (28ft) long and 4.3m (14ft) in diameter, and weighs 2,720kg (6,000lb) before internal fittings are added. At the top of the cylinder, a space has been machined for a 508mm-diameter (20in) viewing port.

## IN THE MOOD

During 1995 the final nails were hammered into the coffins of the space station's congressional enemies. Technical momentum was gathering pace because—at last—NASA had a design they could stick to; and quite apart from paper plans, actual collaboration in space was becoming increasingly routine. On February 3, the shuttle *Discovery* lifted off to rendezvous with Mir, coming to within 13m (37ft) of the station, with astronaut James Wetherbee in command. Eileen Collins sat next to him at the controls, the first female astronaut to pilot a shuttle. As *Discovery* made its final approach, cosmonaut Valeri Titov waved out of the flight deck window to his friends in Mir. A leaking thruster required

**HISTORIC HANDSHAKE**
June 29, 1995: for the first time in 20 years, an American spacecraft commander greets his Russian counterpart in orbit (*left*). Robert 'Hoot' Gibson, in a red tracksuit, shakes hands with Vladimir Dezhurov in the tunnel linking the shuttle *Atlantis* to Mir. Later, inside the spacelab module aboard *Atlantis*, the full complement of crew gathers for a group photo (*above*). Anatoly Solovyev has his arms crossed at the left of the shot. Clockwise from him are Gregory Harbaugh, shuttle commander Robert Gibson, Charles Precourt, Nikolai Budarin, Ellen Baker, Bonnie Dunbar, Norman Thagard, Gennadiy Strekalov and Mir commander Vladimir Dezhurov.

NASA and RSA to negotiate some last-minute revisions to the rendezvous. There were concerns that tiny amounts of fuel might contaminate Mir's delicate external instruments. The eventual solution to this problem set an encouraging precedent.

On March 14, Norman Thagard hitched a ride aboard a Soyuz, to become the first American astronaut aboard Mir. Three months later, the shuttle *Atlantis* came to pick him up, establishing a firm dock with Mir on June 29, 1995 (almost exactly 20 years since Apollo–Soyuz had set the precedent). Two Russians aboard *Atlantis* relieved two others aboard Mir, who came home with Thagard aboard the shuttle. This was the first time in history that people had flown up into space aboard one type of craft, and had come home in another.

## A VISIT FROM MR HANKS

A new Tom Hanks movie was playing that summer, to packed audiences. *Apollo 13* recreated the moon mission of April 1970, which was nearly lost in space as the result of an on-board explosion. Hanks played Jim Lovell, the ship's commander. The interior sequences were photographed aboard a jet cargo plane at high altitude, so that brief moments of weightlessness could be achieved during steep dives. The Apollo cabin, built inside the cargo plane, was a faithful reproduction of the 1970 original, down to the last button and dial. Hanks and his fellow actors gave plausible, low-key performances. Other actors played the ground controllers with equal attention to realism. The great dangers faced by Apollo 13's crew put the successes of the so-called 'routine' moon landings into sharp focus.

The film gripped the public imagination, serving as a potent reminder of what NASA had done in their glory days. A special screening was set up for Bill Clinton in the White House, while the Republicans in Congress claimed the film as an exemplar of their own core values. With this surge of space romanticism backed up by the real-life shuttle/Mir docking, many of NASA's opponents had to rethink their positions if they were not to appear unpatriotic.

On July 25, 1995, Tom Hanks visited Capitol Hill and spoke about *Apollo 13* to a group of politicians from both parties. A few days later he wrote an open letter to Congress: 'I know that such concepts as a permanently manned orbiting science station and other NASA projects are not as glamorous as going to the moon. And Lord knows that our problems here on our world need our attention, resolve and service. But to choose not to go into space, to decide that our days of discovery are over … would hamper our manifest destiny. I hope you will support full funding for NASA's programs.' Hanks's influence and *Apollo 13*'s general effect on the space station political process during 1995 cannot be quantified, but neither can it be discounted. (NASA were pleased at how the movie turned out, though they did wonder how an actor pretending to be an astronaut seemed to have more clout with Congress than real astronauts.)

**HARD DOCK**
A wide-angle view of Mir from the cargo bay of *Atlantis* during the third joint docking in March 1996. The orange-coloured adapter stays attached to Mir.

With a double-Oscar-winning actor on their side, NASA's space station lobbyists were on a roll. A congressional motion to cancel the project failed decisively by 287 votes to 132, at last killing the killers. Bearing in mind the station's one-vote salvation from oblivion just two years earlier, this was something of an improvement. Then, on September 28, 1995, Congress promoted a positive space station bill, championed by Republican representative Bob Walker with support from both sides of the House. Goldin had warned the politicians, 'Our international partners keep getting mixed messages. We have to reassure them that America is serious about the space station.' He also wanted NASA to be assured of long-term funding for Alpha, at a level of $2.1 billion a year between 1996 and 2002, when the last station element would be in place. The endless quibbling over the budget, year by year, was counterproductive and wasteful, Goldin's supporters agreed. The 'Multi-Year Authorization Bill' was passed without dissent. In fact, no formal count of votes was deemed necessary. The result was logged as a unanimous vote of approval, on the floor of the House of Representatives, for America's contribution towards the International Space Station. The complexities of American politics are such that even this vote was no absolute guarantee of future funding stability, but it would be hard for Congress to renege.

Just as NASA celebrated this victory at home, it dawned on them that an outstanding hurdle yet remained. Were the Europeans, the Japanese and Canadians still in the game after all these years of frustration? If not, then the whole deal was off. Dan Goldin had to sweat out one more big vote, far across the Atlantic.

## ESA'S ILLS

On October 18, 1995, just days after the vote in America's House of Congress, ESA's 14-strong Council of Ministers assembled in the French city of Toulouse. The matter in hand was quite simple: would they, or would they not, proceed with their space station contributions?

Quite apart from America's unpredictability as a partner, and the confusions imposed by Russia's involvement, ESA had their own battles to resolve. Italy had some serious budget problems, and was concerned that its contributions to the European collective space effort were not sufficiently recognized in terms of contracts for its domestic aerospace companies. It would only agree to the Columbus space station module and other related equipment if it could build some major components. France was being accused of dragging its feet on major funding issues, and felt it was being asked to contribute more than its fair share towards the space station. Germany was anxious to claim a more significant role. Britain wanted across-the-board reductions in ESA spending. Efforts to design an independent ACRV were getting nowhere. The new Ariane V superbooster needed more funding. French citizens apparently supported the space station idea, but their politicians did not; German politicians were keen, but had no popular mandate.



ESA's member states were pulling in different directions, in part because of the enormous strains imposed on them by the expensive and so far unresolved space station issues. The sudden emergence of a cheap and already operational alternative from Russia proved tempting. France had made a deal to send its own astronauts up to Mir, and was threatening to pull out of the Columbus program altogether. Meanwhile, Germany had booked a satellite launch on a Chinese booster instead of on the Ariane. (Chinese rockets turned out to be explosively unreliable, a bad investment.) The 'Free Flying Platform' element of Columbus had fallen through, due to lack of funding, as had the Hermes spaceplane, after some years of intensive development. The main Columbus pressurized module planned for connection to the space station had shrunk to about half the originally intended size. Many of Europe's programs were under strain, and a political and financial consensus was vital at this late stage. With Russia now aboard and Congress tamed, the last thing Dan Goldin needed was for ESA to fall apart on him at the eleventh hour.

The positive messages from America, the luxurious surroundings of Toulouse's 17th-century Town Hall, and plenty of rich French food and wine set the tone for a meeting that was far more productive than the Europeans had dared hope for. ESA chiefs pleaded with the member states to hold the fort. The spectre of European humiliation in front of America and Russia proved too awful to contemplate, and the main Columbus space station module was approved, along with development of an Automatic Transfer Vehicle, a cargo supply pod similar in principle to Russia's Progress ferry, to go atop the emerging Ariane V. Contracts were divided on an equitable basis, and the manned research program was restarted, in the form of modest studies for an escape capsule, and with the comforting knowledge that Soyuz capsules would take up the slack, thus allowing ESA plenty of time to work on developing their own vehicles. Even Britain was caught up in the mood of renewed optimism, withdrawing its demand for a 25 per cent reduction in space science spending and offering 2 per cent towards continuing development of the new Ariane V.

James Beggs's gamble had finally paid off.

## RUSSIAN ROULETTE

In October 1995 the respected aerospace magazine *Flight International* was trumpeting all these agreements in bold headlines: 'Space Station Funding Passes Major Milestone' and 'Europeans Resolve Space Station Row'. By December 1996, the news seemed less encouraging: 'NASA Concerned Over Russia's Commitment. Space Station Faces New Crisis.' At the time, a frightening disintegration of the NASA and RSA alliance seemed to be looming on the horizon like an incipient thunderstorm. For many months afterwards, the scare stories created the impression that Russia was an unstable ally. The truth was more subtle.

On the surface, RSA had some legitimate concerns. NASA had agreed to pay $400 million for seven docking visits with Mir, including the use of two additional modules, Spektr and Priroda, which were filled with American experiments and then launched (late, but launched nevertheless) on unmanned Proton boosters during 1995 and 1996. But the

1993 agreements were looking rather one-sided, with America gaining more than Russia in the longer term. Nationalistic politicians in Moscow were not convinced of the deal's benefits. In addition, objections within the military establishment to the new space station precisely mirrored the DoD's traditional opposition to foreign involvement in NASA's plans.

RSA felt some pressure not to scrap Mir after 1997, when the first International Space Station modules were due for launch. Though ageing, Mir still had a few years' life in it, and the modules that had been delivered for the benefit of visiting American astronauts were nearly as good as new. It seemed foolish to scrap them.

Quite apart from FGB, which would be purchased by Boeing under contract for NASA, a brand-new spacecraft was already sitting in the Khrunichev assembly plant at Baikonur, waiting to fly as the core module for Mir's successor. Russia now had little chance of launching this second-generation Mir. As an extension of the September 1993 agreements, it had been decided to adapt this factory-fresh hardware as a 'Service Module' for the new space station, providing initial life support and a crew habitation area complete with shower and toilet. This would constitute Russia's primary hardware contribution. While FGB was being paid for by America, the Russians would have to finance the Service Module themselves, just as ESA and Japan were financing their own pressurized modules.

In December 1995, Russian space officials suddenly warned NASA that their FGB Functional Cargo Block for the new space station could not be delivered on time. Since the FGB was now absolutely critical to NASA's entire schedule, this came as a shock. The RSA then calmly came up with a new deal. Instead of the new Service Module, Mir itself would form the core of the International Space Station. The whole idea was appalling. Russia's plan would involve yet another costly redesign. All the congressional support for which NASA had fought so hard and so long could be destroyed at a stroke, and anyway, this wasn't what RSA had agreed to back in 1993. Boeing had already signed a deal with Khrunichev for a factory-fresh FGB. They didn't want something second-hand and 10 years old. Urgent meetings were arranged, and NASA told the Russians that a renegotiation of their original deal was not an option.

Gradually, American officials began to sense mixed messages coming from the Russian political establishment. While the RSA negotiators badgered NASA with terrible scare stories, two Republican congressmen, James Sensenbrenner and Jerry Lewis, visited Moscow in January 1996, seeking clarification of these problems from senior Russian



**SPACE STATION:**
**THE FINAL VERSION?**

| | |
|---|---|
| **Width** | **110m (360ft)** |
| **Length** | **88m (290ft)** |
| **Weight** | **462 tons** |
| **Altitude** | **335km (220 miles)** |
| **Orbit** | **51.6 degrees** |
| **Crew** | **3 interim, then 6** |

**PRINCIPAL COMPONENTS**

| | |
|---|---|
| **1** | *Solar Power Arrays* |
| **2** | *Thermal Control Radiator* |
| **3** | *Russian Research Module* |
| **4** | *Connecting Node 1* |
| **5** | *Docking & Stowage Module* |
| **6** | *US-built Habitation Module* |
| **7** | *Soyuz Crew Transfer Vehicle* |
| **8** | *Connecting Node 2* |
| **9** | *Japanese Logistics Module* |
| **10** | *Japanese Experiment Module* |
| **11** | *Japanese Exposed Facility* |
| **12** | *US Space Shuttle Orbiter* |
| **13** | *Photovoltaic (PV) Arrays* |
| **14** | *Support Module* |
| **15** | *Russian-built Service Module* |
| **16** | *Attitude Control Gyroscopes* |
| **17** | *Thruster Facility* |
| **18** | *Russian-built FGB* |
| **19** | *Central Truss & Gyros* |
| **20** | *US Laboratory Module* |
| **21** | *Centrifuge Module* |
| **22** | *Canadian Remote Manipulator* |
| **23** | *Prefabricated Truss Elements* |
| **24** | *ESA Columbus Module* |
| **25** | *External Experiments Location* |
| **26** | *PV Array Rotary Joints* |

1
2
3
4
5
6
7
8
9
10
11
12
13
14
15
16
17
18
19
20
21
22
23
24
25
26
NASDA

## THE ASSEMBLY SCHEDULE

Between launch of the first operational space station element in November 1997, and completion in June 2002, RSA and NASA will have to make at least 34 flights between them to bring up all the components. NASA's shuttles will fly 28 assembly missions, with RSA's expendable Protons contributing 10 launches. In addition, Russia may have to send up more than 20 Soyuz and Progress supply ferries on slightly smaller boosters, at the rate of 5 a year. Japan's scientific module will hitch a ride on a NASA shuttle during March 2000. ESA's own launch vehicle, the Ariane V, will lift the Columbus laboratory module into orbit during September 2001.

Assembly flights will bring up a tremendous range of equipment: navigation gyros, solar power cells, internal and external experiments racks, truss structure elements, batteries, airlocks, robot manipulators and docking collars. Major milestone targets center around delivery of the following large components:

| Date | Agency | Payload | Purpose |
|---|---|---|---|
| Nov 1997 | RSA | FGB | *Functional Cargo Block, provides interim power, docking, attitude control and propulsion capability* |
| Dec 1997 | NASA | NODE 1 | *Initial US pressurized element, provides interface with Russian hardware, and a shuttle docking location* |
| Apr 1998 | RSA | SERVICE MODULE | *Initial life-support capability, plus docking collar for Progress ferries* |
| May 1998 | RSA | SOYUZ | *Escape capability for interim crew of 3* |
| Sep 1998 | NASA | SOLAR ARRAY | *First array, installed in a temporary location above Node 1* |
| Nov 1998 | NASA | US LAB | *Main pressurized US scientific module* |
| Dec 1998 | NASA | RMS | *Canadian Remote Manipulator System to assist with station servicing* |
| Mar 1999 | NASA | AIRLOCK | *Joint spacewalk capability for RSA/NASA astronauts with common spacesuits* |
| Mar 2000 | NASA | JEM | *Japanese Experiment Module* |
| Aug 2001 | NASA | CENTRIFUGE | *Scientific equipment, simulates range of gravity conditions* |
| Sep 2001 | ESA | APM | *Attached Pressurized Module Columbus* |
| Jan 2002 | NASA | SOLAR ARRAY | *Final solar cell panels installed in permanent positions* |
| Feb 2002 | NASA | US HAB | *Habitation Module for entire crew* |
| Feb 2002 | RSA | LIFE-SUPPORT MODULE | *Oxygen regeneration and other support* |
| Jun 2002 | RSA | SOYUZ | *Second escape system, enabling full crew of six to remain aboard the station* |

politicians. They were told there were no problems. The RSA space station components would be delivered on time, just as agreed. Please don't worry, the politicians said. Somewhat confused, Sensenbrenner and Lewis headed for home. And then it dawned on them what had happened. As one of their aides later explained, 'The RSA wasn't getting as much political cooperation in their own country as they wanted, so they came here, scared the hell out of us, and got us to put pressure on the Kremlin.' Just as Beggs had gambled on matters of foreign prestige to swing key votes in Congress, so the RSA reinforced faltering political and financial support at home by effectively threatening their own government masters with embarrassment. President Yeltsin and his colleagues were not about to upset deals that brought millions of dollars into their country, and RSA used this to their advantage.

By the end of January 1996 a new agreement was on the table, one that seemed quite reasonable. In return for the RSA's continuing commitment to the space station, NASA shuttles would take up some of the burdens of replenishing the current Mir, thus freeing up Protons for other work. In addition, six space station assembly launches would be struck off the RSA flight rota, and their piecemeal payloads of solar power panels and other equipment would be taken up instead on shuttle flights. NASA could afford to do this, partially by reshuffling their list of future launches, but they had made some significant concessions nevertheless. In a strange way, it was an encouraging development, proving that RSA was learning how to play—and win—the game of space politics. NASA seemed so relieved at saving the broader alliance, they hardly noticed how much ground they had surrendered.

Meanwhile, Lockheed, NPO Energia and Khrunichev were collaborating to exploit Proton's potential as a commercial satellite launcher. By April 1996 this had borne fruit, with a successful launch of a Belgian communications satellite. The harsh restrictions of the 1993 commercial launch agreements with Russia will no doubt be eased in coming years if companies like Lockheed risk losing out because of them. American policies aimed at protecting the domestic rocket industry obviously have to be adapted if those same companies are putting money into Protons and other Russian boosters. Protectionism will also backfire against Western companies who provide payloads. Hughes, Inmarsat and Motorola have already booked 11 Protons between them for communications satellites.

With deals like these, Russia's space effort now has the faint traces of a 'hardball' negotiating power, albeit arranged largely with the help of foreign marketing experts. These can only be good developments, both for Russia and for their sudden profusion of international partners.

NASA

**GETTING ALONG**
June 1995: Norman Thagard during his three-month stay aboard the Russian Mir, with Gennadiy Strekalov (*right*). Thagard was made welcome aboard Mir, but on occasion he felt culturally isolated.

# THE HUMAN FACTOR

*With so many political and technical problems solved, space station planners must face up to the wayward human element. Aboard the new station, crew members from different cultures will have to tolerate each other for long periods in difficult conditions.*

**NO WALK IN THE PARK**
Space walking looks relaxing to the untrained observer, but astronauts can find it gruelling and painful.

# THE HUMAN FACTOR

**CLUMSY COVERING**
Spacesuits become rigid when inflated with air, like expensive balloon animals. 'Constant volume' rotary or bellows joints at the wrists, elbows, shoulders and knees do not change shape or lose flexibility when the suit is pressurized, and their design allows the wearer some degree of mobility, but delicate movements are difficult. The photo shows a typical space shuttle suit design from 1988, stripped of the deceptively flexible-looking outer white cladding of micrometeoroid and thermal blankets. Gloves have always posed a problem. They need to be flexible for delicate work, but they are the most likely part of a suit to become punctured, so they have to be sturdy. Astronauts' fingers can become blistered, and the effort required in just bending the glove's fingers can leave the hands weak and numb. Astronaut Kathy Thornton once described working in EVA as 'like fixing a car engine while wearing boxing gloves'.

American space station engineers are quietly confident that their machinery will work. Most of the critical assembly techniques, from welding the modules on the ground to docking them together in orbit, fall well within their established range of expertise. It is only once the station is up and running, with people aboard, that NASA will really come up against a barrier of inexperience. Their practical fund of knowledge about living and working in a space station is based largely on the Skylab program, which ended in 1974. But time has not diminished the importance of the lessons learned.

Two Skylab veterans, Jerry Carr and Bill Pogue, are now employed by Boeing as consultants. In advanced middle age, they are still trim and fit enough to clamber into space suits and practise station procedures in the giant watertank at Marshall. After more than a quarter of a century, the rewards of life on Skylab remain vividly fresh in their minds—as do the countless irritations. They have claimed a special academic discipline as their own: how *not* to design a space station.

Pogue and Carr, along with crewmate Ed Gibson, were the third and last Skylab crew. When they clambered on board during November 1973 for an 84-day tour of duty, expectations were running high. The preceding two crews had worked well, repairing the damaged station and throwing themselves into an intensive work routine. But on day one, the third crew started off on the wrong foot. Pogue threw up (or 'out') almost as soon as he entered the station's cavernous—and weightless—work area. Unwilling to show him up, mission commander Carr said, 'We won't mention that. We'll just throw the mess down the trash airlock.' But later, mission control reprimanded the crew. Their private conversations were tape-recorded just as thoroughly as any in the Nixon White House.

There followed several weeks of intermittent tension. Skylab's preceding two crews had set a brisk pace at work. The third crew felt they were being pushed too hard. NASA's ground controllers tried to fit in the endless requests from scientists to relay instructions up to Skylab. The requests were neatly tabulated, organized, *managed*, at ground level. It was very different when Carr, Gibson and Pogue actually tried to put these requests into action. The astronauts' general response was—if you think you're so damn' clever down there, you try doing it here in Skylab. The crews' frustrations were not eased by Skylab's general layout. Engineers had used floors, ceilings and walls to mount equipment, control panels and storage lockers, so that every available area of interior space could be put to good use. The crew's sense of distance became warped, because the horizon that guides us in our daily lives was not apparent in some of their enclosed work areas. In addition the usual differences in light levels, brighter above our heads, darker at our feet, were nullified on Skylab by the poor placement of light fittings, sometimes on the ceiling and sometimes on the walls. As a result, the astronauts hardly knew from one moment to the next which way around they were. 'We really need a better sense of up and down with a proper difference between the floor and ceiling,' Pogue fretted.

Nor was the decoration scheme to their liking. 'The colour scheme in here has been designed with no imagination. All we've got in here is about two shades of brown, and that's it for the whole damn spacecraft interior,' Gibson complained. The crew's carefully fireproofed garments were as drab as the wall colours, as well as feeling stiff and prickly. They yearned for floppy tee-shirts and more casual, comfortable attire.

These were not the least of Skylab's drawbacks. The bathroom was unpleasant to use. The metal floor was designed as a hygienic wipeable surface; it had no firm footholds, and at washtimes the astronauts slithered about in there like sardines in a can. Having a good wash took up not minutes but hours, because droplets that strayed from the plastic shower shroud had to be hunted down and mopped up before they could seep into any electrical equipment and short-circuit the works. These hours tended to cut into precious leisure time. As Carr joked bitterly, 'Off-duty activities? You gotta be kidding! There's no such thing up here. On a day off, the only difference is you get to take a shower.' The shaving mirror was made out of dull metal, to eliminate the risk of injury from broken glass. The crew never got a good look at their own faces, an apparently minor inconvenience that turned out to be strangely unsettling over time.

And there was more. The storage locker numbering system made no sense whatever, so the crew could never find important tools, or anything at all, for that matter. When they did find the right locker, everything inside it would tumble out as soon as the door was opened, and then all these silly little items that they didn't want just yet—pens, screwdrivers,

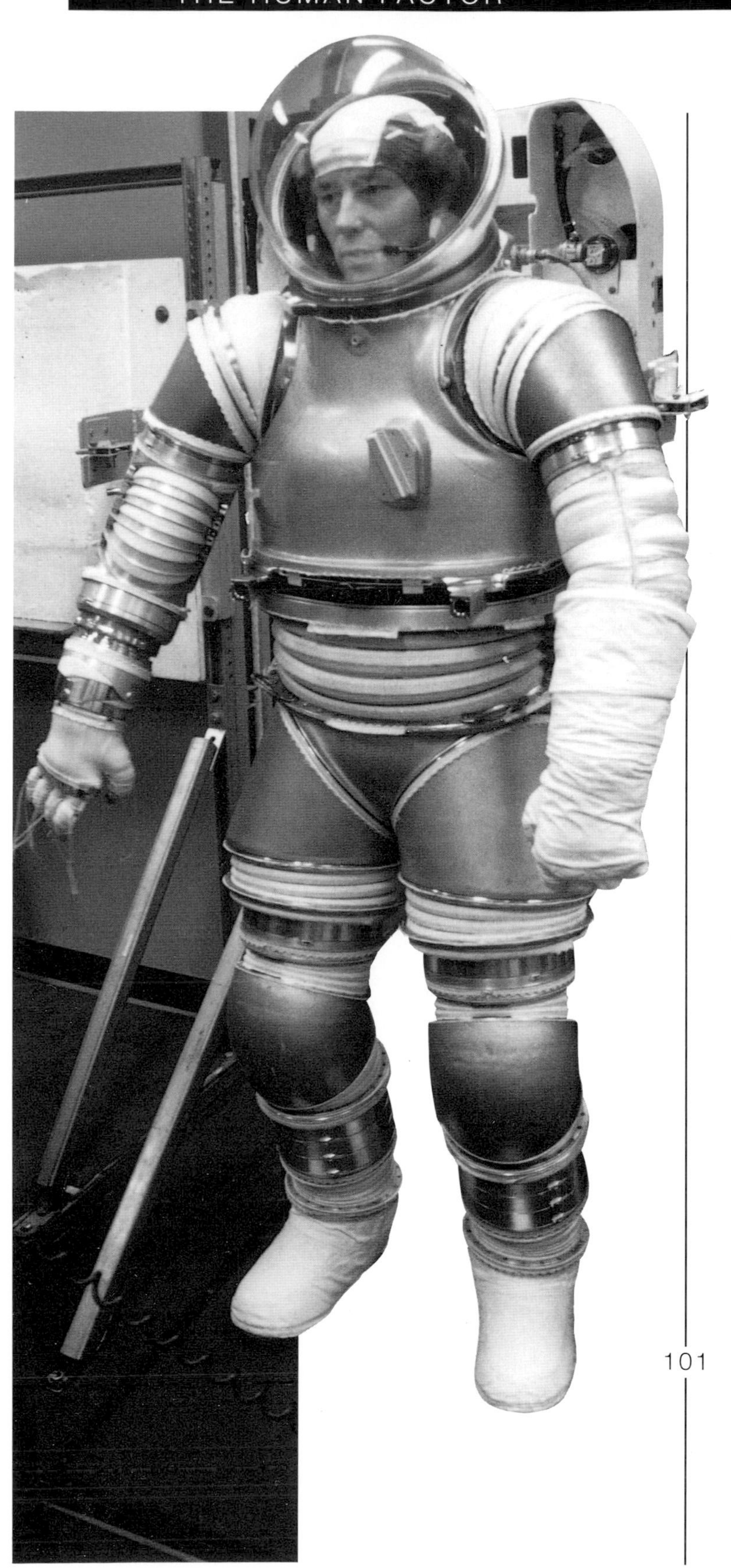

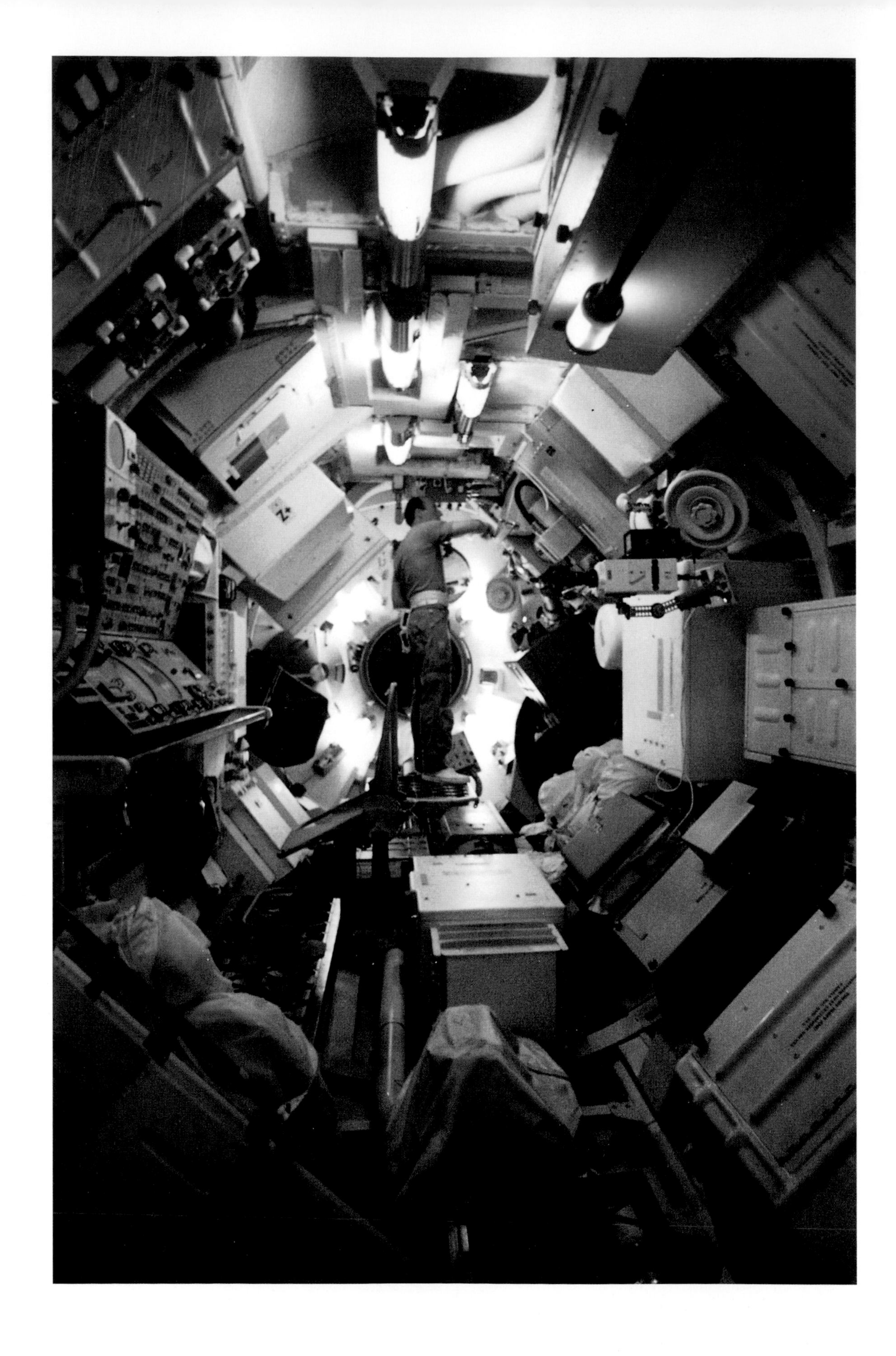

cloths, notepads, film cartridges—would have to be retrieved and stuffed back inside. Soon the entire station would fill with drifting junk, and the crew had to watch out not to choke themselves by breathing something in: a lost screw, perhaps, or a stray shred of paper.

The best bet was to wait for a few hours, when all the loose junk would eventually get caught by the gentle suction of the air-conditioning inlets, which then had to be cleaned out by carefully removing dust and stray food crumbs from the grilles with a small vacuum hose. This would lead to a row with mission control because the useless managers down there hadn't scheduled this unforseen but important task into the astronauts' busy day. Petty irritations like these could make life on Skylab unbearable at times, and also led to inefficiency. At one point Bill Pogue was struggling with a scientific package: 'There's no way we can do a professional job. I don't like being put in a position where I'm taking somebody's expensive equipment and thrashing about wildly, trying to get it started in insufficient time!'

Worst of all, the crew's tight work schedule prevented them from appreciating the actual experience of being in space—the search for which had driven their astronautic careers in the first place. Every spare moment, they huddled around Skylab's single small window, staring in childlike fascination at their home world, seeing it as very few people before or since have ever had a chance to see it.

**NOT LIKE YOUR MOTHER'S HOUSE**
The 3.05m (10ft) diameter docking and equipment module on Skylab was the least satisfactory work area, with no proper sense of up and down, and sharp-edged boxes of equipment protruding from every available surface. The big grey switch panel on the left of this picture housed attitude thruster controls, gyro readouts and solar telescope screens. The crew would have liked a friendlier working environment. At least Skylab was *big*. The diagram of the new space station as compared to Mir (*right*) shows that for all its size, most of the station's bulk is made up from superstructure elements. The work and habitation modules will still be fairly cramped.

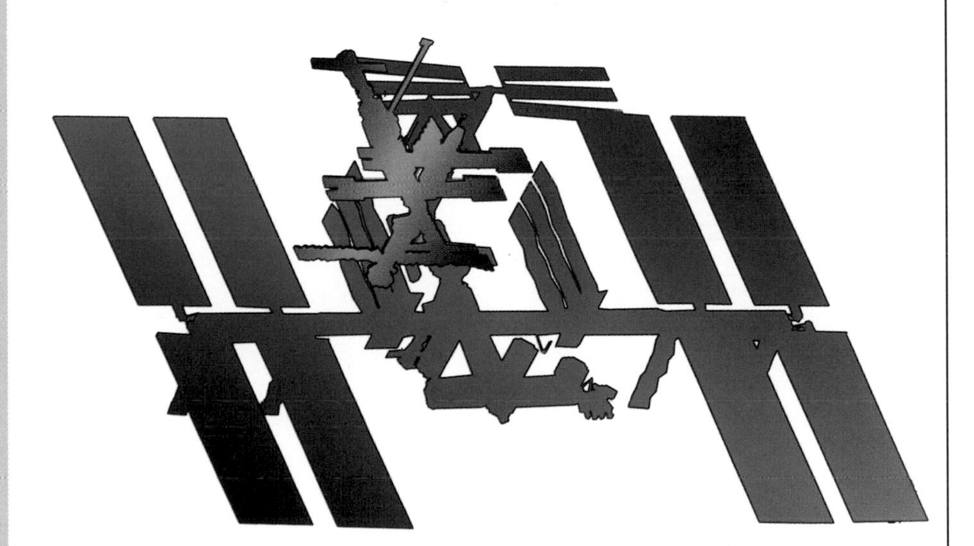

Frustrations aside, Pogue, Gibson and Carr put in an astounding work performance, and derived a great deal of professional satisfaction from their 84-day flight. Keeping in mind that this was still a very new experience for everybody at NASA, the occasional complaints were seen as valuable learning tools, both then and now. Skylab's engineers quickly realized that their designs, although superbly suited to the challenge of packing all the complex machinery on board, had fallen far short of making a proper home in space fit to live in. Today, Boeing and the other station contractors are keen not to make the same mistakes. But no matter how many sophisticated psychological and ergonomic theories they bring to bear on these problems, certain aspects of design come down to nothing more than personal taste. In 1991, NASA's consultants somehow determined that pink was a restful colour suitable for a station's living quarters. Shuttle astronaut Joe Allen dismissed this notion: 'I don't live in a pink house. Why would I want to be in a pink spaceship?'

## NEW AND IMPROVED

With Skylab in mind, the new space station will not be so topsy-turvy in its layout. The colour scheme will be brighter and more cheerful, with an inherent bias towards 'up' and 'down'. The ceilings will contain strip lighting, while the floors will have dark-coloured panels to differentiate them. Airlocks and hatchways will be marked with a 'top' and 'bottom'. The general interior arrangement will be much less ragged than in the past, with awkward boxes of equipment neatly tucked away behind a relatively smooth facade of access panels. Crews will be able to move swiftly through the station without banging their shins against protruding sharp edges.

Much of the Skylab crews' work consisted of using, and just as often fixing, complex machinery. On the new station all the scientific and operational equipment will be mounted in hinged racks so that machinery behind the front panels can be easily reached during maintenance without undoing countless screws. Racks of redundant equipment will be easily replaceable with new racks brought up on a shuttle. During repair of the more delicate items of gear, astronauts will make use of a special work bench where a gentle downdraft of air against a fine screen will prevent small objects from flying off. Storage lockers for all the spare parts and ancillary gear will be sensibly and clearly marked, and the contents firmly secured so as not to come leaping out every time the lockers are opened.

One by one, the potentially dangerous irritations recounted by Carr and Pogue have been chased down and eliminated. The practical difficulties of life in space become exaggerated to an unholy degree by the unusual environment. Finding out about these factors is part of what a space station is for. At least, it is now. In the Skylab days, most NASA crew members had backgrounds as fighter pilots. They were a tough, competitive species unaccustomed to making complaints about their comfort levels, partly because they were never given the opportunity. Pogue recalls, 'NASA didn't downplay the stress factors. It ignored them.' Today's planners recognize that good design, in space as on earth, is essential in creating a pleasant and stress-free environment.

Meanwhile, the greatest pleasure of life on Skylab will be enhanced for inclusion on the new station. Windows will be large and comfortably placed. In addition there will be a special viewing cupola with all-round vision attached to one of the modules. This will provide excellent visibility for docking and maintenance work, for supervising space walks, conducting earth photography and so forth—but its secondary purpose will be no less important: allowing the station's crew an unprecedented

**IS SPACE BAD FOR YOU?**
July 7, 1995: Norman Thagard back aboard shuttle *Atlantis* after his stay on Mir. Despite a vigorous exercise regime throughout his mission, Thagard's muscle tone has deteriorated, especially in his legs. Loss of bone calcium and a host of other medical problems result from weightlessness. Back on earth, Thagard's body will quickly recover, but nobody yet knows what the long-term effects of extended space flights may be. However, the study of such medical phenomena may well justify sending people into space in the first place.

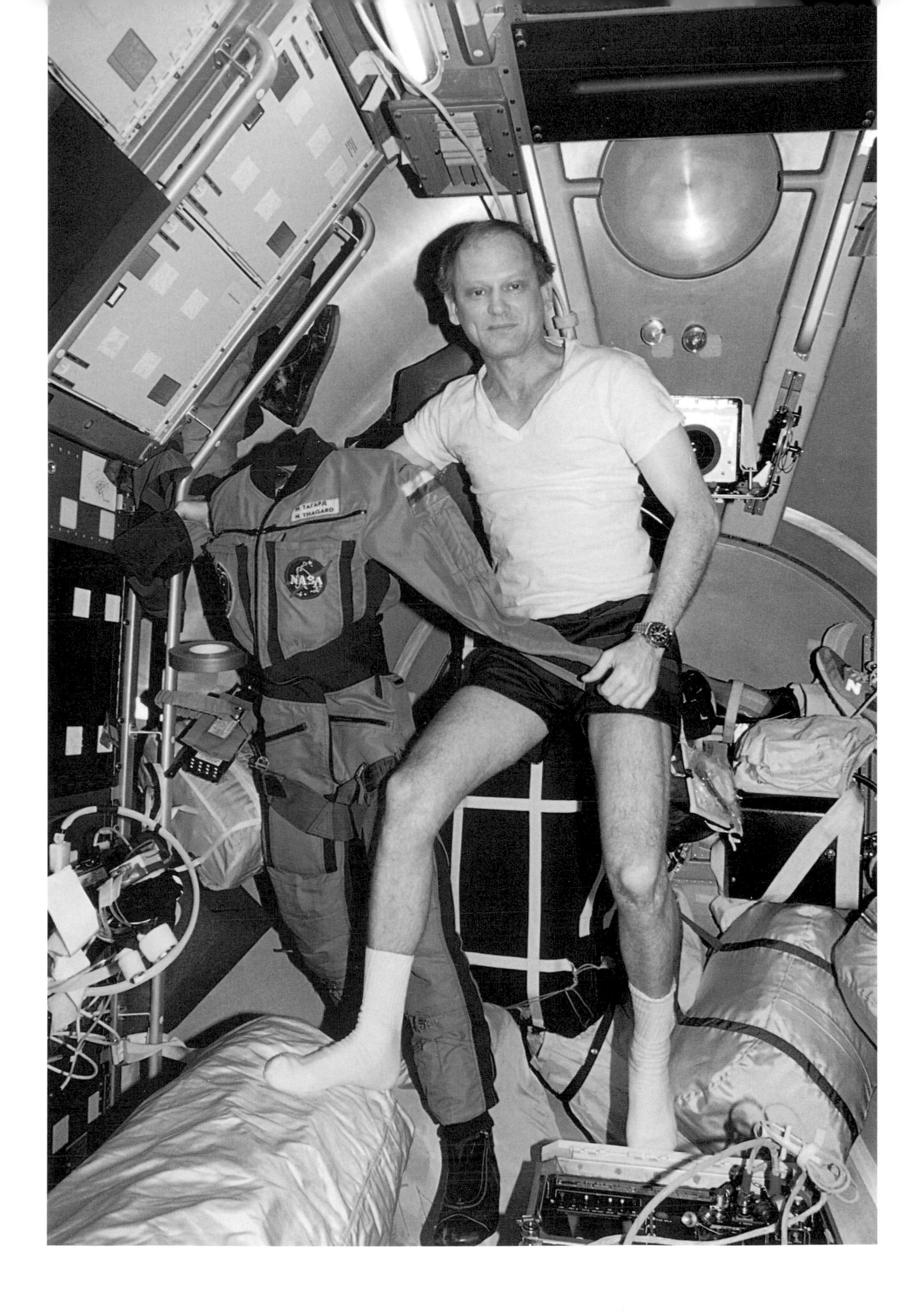
N. THAGARD
NASA

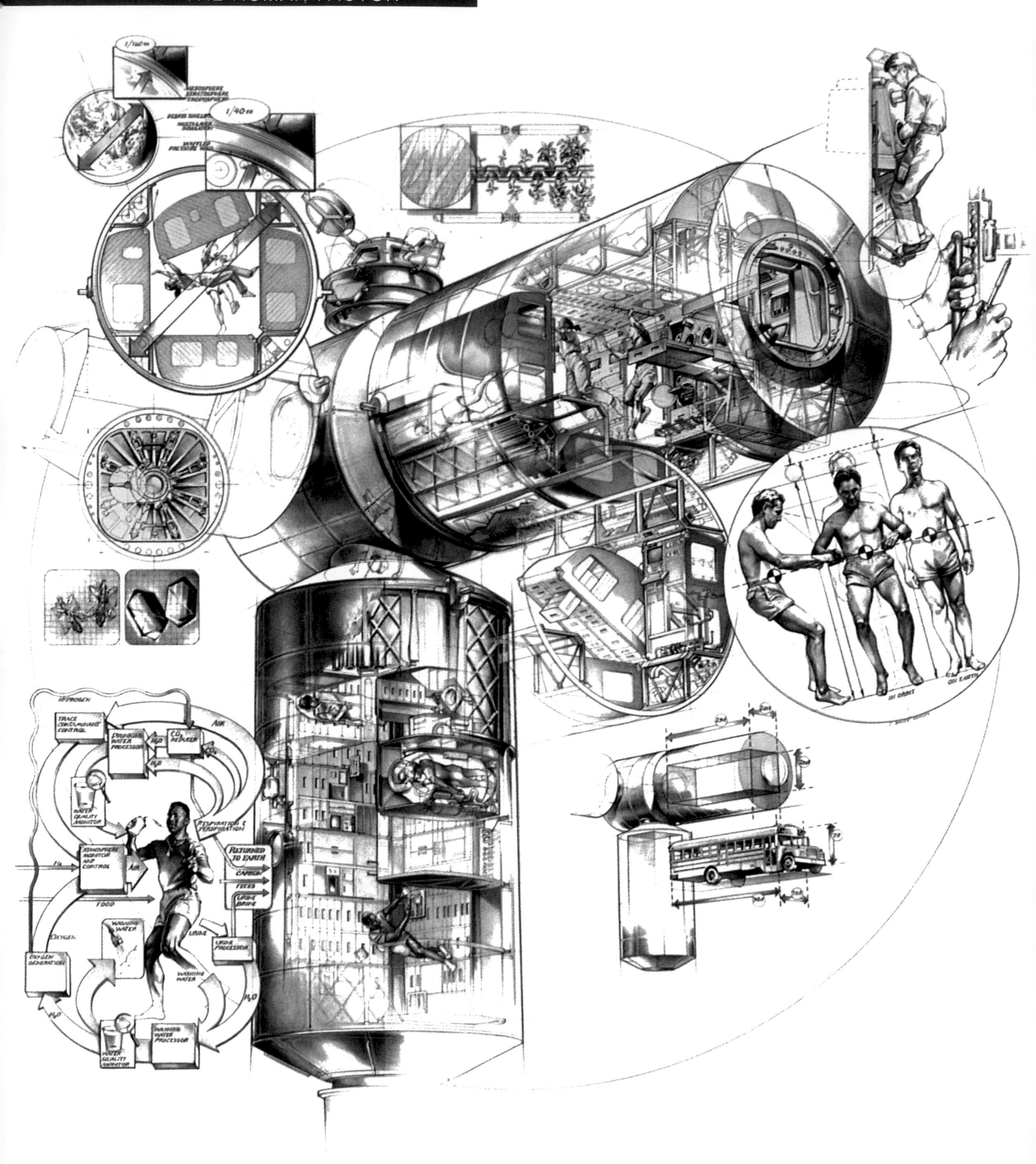
MESOSPHERE
STRATOSPHERE
TROPOSPHERE
DEBRIS SHIELD
1/40 in
HYDROGEN
TRACE CONTAMINANT CONTROL
DRINKING WATER PROCESSOR
$H_2O$
$CO_2$ REDUCER
$CO_2$
AIR
WATER QUALITY MONITOR
RESPIRATION & PERSPIRATION
ATMOSPHERE MONITOR AND CONTROL
AIR
RETURNED TO EARTH
CARBON
FECES
URINE BRINE
FOOD
WASHING WATER
OXYGEN
URINE
URINE PROCESSOR
OXYGEN GENERATION
WASHING WATER
$H_2O$
$H_2O$
WATER QUALITY MONITOR
WASHING WATER PROCESSOR
IN ORBIT
ON EARTH

opportunity to gaze at the scenery in their leisure periods. The earth turning gently beneath them will provide perhaps the most wonderful view any human can experience, and also one of the most poignant.

## LOST IN SPACE

A space station never flies more than a few hundred kilometres above the ground. It remains so close to earth that atmospheric drag is a constant problem, requiring the station to be reboosted at intervals if it is not (like so many Salyuts, and even Skylab) eventually to slow down and fall back to earth. This proximity to the ground enables shuttles and ferries to make a rendezvous without burning up absurd amounts of fuel. In principle it can take less time to come up and link with a station than to fly across the Atlantic from New York to London. But as Russian cosmonauts aboard Mir have discovered, these physical distances can be dwarfed by the psychological chasm that opens up between an orbiting space crew and their home world.

**SCIENCE AND HABITATION MODULES**
This Boeing publicity artwork shows the internal structure of the American laboratory and habitation modules for the International Space Station.

A few days' disconnection from earth can be tolerable, but weeks or months of it can be a strain. So far, it has been the long-duration cosmonauts who have suffered the most. It is tempting to say there is more poetry in the native Russian soul than in the American. Mir's crew members often experience a sense of melancholy, an imaginative understanding of what it really means to be adrift in the cosmos. Even as they grit their teeth and pursue their incredibly resilient routines in space, they long for home, spouse, family.

Their mood is hardly improved by the fragility of their radio link with the ground. Mission controllers in Kaliningrad (a suburb of Moscow) can only maintain intermittent contact with Mir. In the 1970s Russian radio-relay ships plied the high seas, their giant dish antennae straining at the skies to catch every passing cosmonaut's quietest thought. Constant communication was thus assured. For much of Mir's career, these ships have not found sufficient funding to put to sea. Consequently, Mir can only keep in touch when it is directly over Russian territory.

Sergei Krikalev's experiences in 1991–2 must have been particularly poignant, with occasional, unsteady radio transmissions telling him, for 15 minutes at a time, of great and unspecified upheavals down on the ground, before Mir's orbit swept him out into long hours of silence.

Not *that* much silence, however. Space stations are very noisy places. The big problem is getting air to circulate properly. On the ground, convection moves cold air down, warm air up (because warm air is lighter, less dense, and it floats to the top of the cool air). Ordinary air-conditioning systems use this simple principle to change stale air for fresh, pumping out cold fresh air, which we breathe at floor level till it warms up and drifts towards the ceiling vents, to be sucked away and pumped off the roof of the building. In the weightless space environment, convection simply doesn't happen. A sleeping astronaut can be suffocated by a smog of her own warm, stale exhalations, and no natural forces will clean the air for her unless a current is artificially generated. Space station atmospheric systems rely on brute power to shift air around. They also

require intakes and outlets scattered widely around all the cabins, and not just discreetly tucked into floor skirtings or ceiling panels. A colossal amount of electrical power feeds into the many electric fans, which shudder and shake inside the morass of pipes and ducts. Drafts become a constant problem, chilling the backs of astronauts' necks and drying their mouths. Nor is there any question of simply pumping stale air overboard. Big noisy boxes called 'scrubbers' recycle old air, removing carbon dioxide (using lithium hydroxide filters) and reinforcing the breathable content so that as little oxygen as possible is wasted. Condensation and fungal build-up in the scrubber vents have created constant problems in Mir. Just to round off the whole fragrant ensemble, urine is also recycled within the environmental control systems to remove its precious water content for drinking and washing, and sometimes this process also falls prey to dangerous impurities.

## CLOSING THE LOOP

Spacecraft designers have tried frequently to design self-sufficient 'closed-loop' environmental control systems, in which all air and water is recycled, and food too, as far as possible. None of these designs has yet proved acceptable for use in space. Missions to Mars will need very efficient control systems, unless the whole vast tonnage of the ship is taken up with food, air and water; but so far, completely self-supporting life systems remain elusive. The new space station will rely heavily on replenishment missions from earth. Some commentators have argued that space planners have no right to expect that they can come up with a truly closed-loop environment for a space station when they can barely understand the parallel mechanisms at work in nature.

The argument works both ways. Life-support experiences gained aboard the station may contribute new insights on the broader question of how air and water are recycled in the earth's biosphere.

Then there are the instruments and scientific experiments that are a space station's whole purpose—all of which need their own comfortable working conditions. Just like ordinary office equipment, these systems require fans to keep their electronic circuits cool; but with no convection to help the process, these fans have to be more numerous and more powerful than usual. Skylab was noisy enough. Mir, drifting in the utter silence of the cosmos, is literally deafening. Cosmonauts sometimes come home with their hearing permanently damaged.

**CLEAN CABIN**
In comparison to Skylab or Mir, the International Space Station will provide very orderly working areas, with no harsh surfaces or unwelcome clutter. However, the neat interiors as designed will probably develop a lived-in look over time.

**SPACE CONSTRUCTORS**
The McDonnell Douglas artwork at left shows astronauts on the space station's central truss element, which attaches the modules to the general superstructure. Highlighted in blue is the Canadian Remote Manipulator System, which can move from side to side along the truss. Astronauts are now training for space station assembly tasks in the water-tank facility at Johnson Space Center (*right*). Some of the old Freedom designs called for perhaps 2,000 hours of spacewalks, 'Extra-Vehicular Activity' (EVA). Bearing in mind that NASA's total EVA time, including Skylab and the Apollo moon walks, has amounted to barely 600 hours, and Russia's, 260, this requirement has always been undesirable. As things stand, the space station will need 650 hours of EVA time between 1997 and 2002, divided between Russian and American astronauts, with an extra 170 hours per year for routine maintenance.

## DISUNITED STATES IN SPACE?

The noise, the smell, the stale water, the colour scheme, the crazy design of the equipment: all of these can be tolerated so long as you have friendly human company. But long stays aloft can create friction between the closest teams. In 1978, a Salyut crew had to be brought home early because one of them suffered a nervous breakdown. Mir's crews have often bickered, sometimes with each other and more often with mission control (a traditional target for all space fliers' frustrations). In early 1995, prior to the first shuttle visits to Mir, cosmonaut Gennadiy Strekalov made a number of spacewalks to move solar panels and other equipment on the exterior of the station, so that the shuttle would have a clear approach path. As the spacewalk schedules became more intensive and, in Strekalov's opinion, increasingly haphazard, he rebelled. 'These spacewalks have not been properly rehearsed. They are dangerous. I won't do them!' Eventually, Strekalov relented and carried out his work flawlessly, but on his return to earth he was fined a substantial proportion of his salary.

Occasional tensions between crew members are usually offset by a common bond of training, loyalty to mutual employers and a shared cultural background. A multinational space station crew will face unprecedented difficulties. Norman Thagard, the first American astronaut to spend time aboard Mir, found the cultural isolation hard to bear. The intermittent communications with ground stations meant that sometimes he had to wait two or three days at a stretch before hearing a familiar American voice for a few precious minutes. Thagard had spent many weeks in Star City, the cosmonaut training complex near Moscow, and had learned to handle all the Russian machinery, space suits, valves and

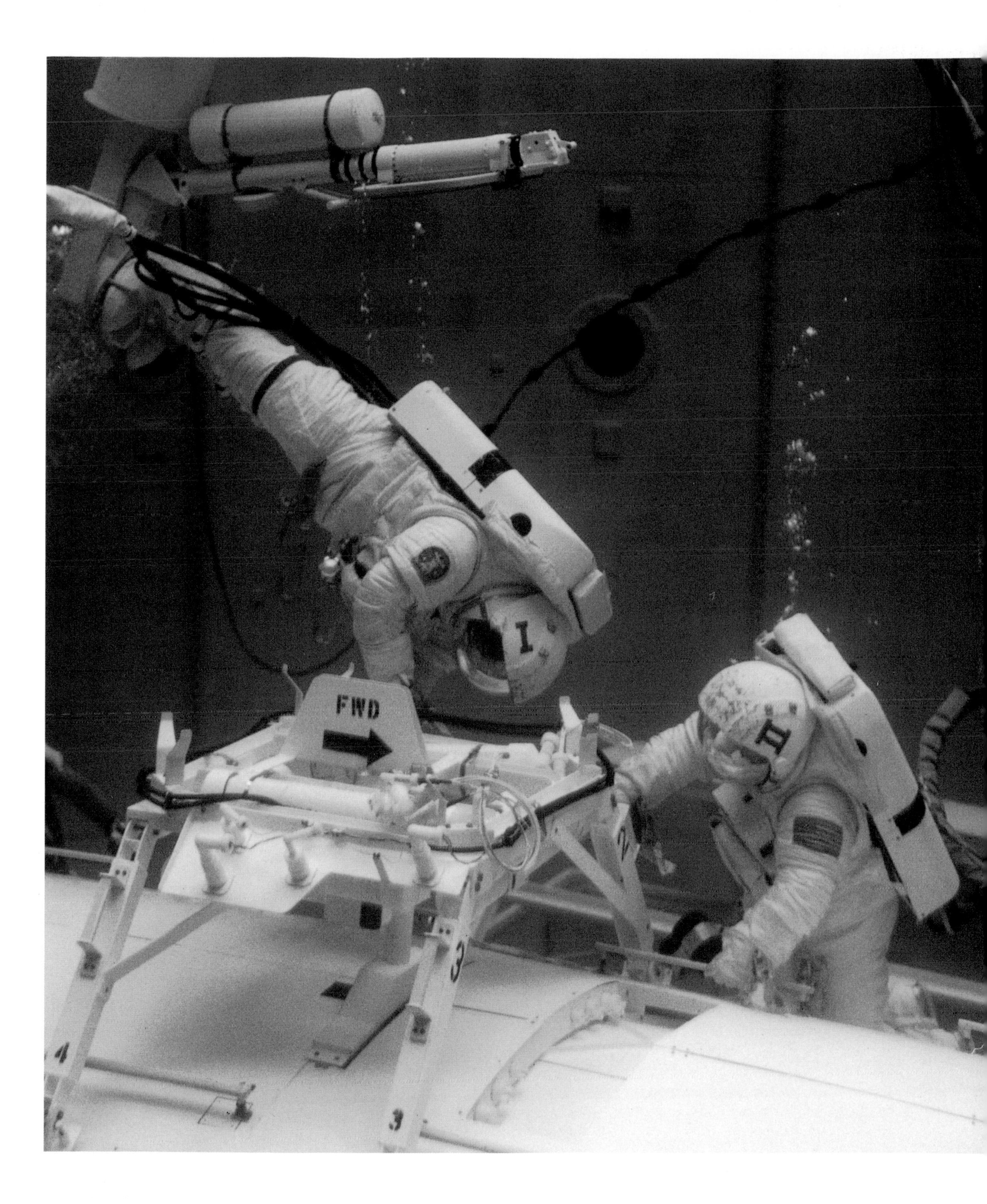
FWD
I
II
3
4
3

airlocks. Once in orbit, the psychological isolation took him completely by surprise. His Russian was good enough to make conversation, but not so good that he could feel close with his crewmates or crack jokes with them. Somehow, his training had not prepared him to deal with Russian people.

With these kind of problems in mind, McDonnell Douglas, the company partially responsible for equipping the interior of the new space station's six-berth habitation module, was the first NASA contractor ever to hire an anthropologist to study the interactions of people in space. Dr Mary Lozano interviewed astronauts from America, Europe, Japan and Russia, and discovered significant cultural preconceptions. 'We know people can get on each other's nerves, specially in what we call a "trapped environment". Imagine the problems that occur if you come from widely-differing cultures.'

Dr Lozano's Japanese interviewees, for instance, felt that American astronauts and mission controllers tended towards making rapid judgements without consulting their superiors. NASA's flight staff, often called upon to make split-second decisions that might be crucial to the safety of a mission, thought that the Japanese preference for group decision-making could be fatal in an emergency. Americans felt that the Japanese tended to say 'yes' when they meant 'no', because of their anxiety to maintain social harmony. To the Japanese mind, this individualistic approach seemed dangerous and irrational. They thought of Americans as arrogant. Americans considered the Japanese clannish, and the French arrogant. Germans thought the Italians over-emotional. Italians found that Americans had less sense of personal privacy. Canadians were annoyed at being thought of by Europeans as 'just the same' as Americans.

Dr Lozano also found that Russian cosmonauts also thought of NASA astronauts as too individualistic, too overtly competitive. These particular studies for McDonnell were conducted in 1992–3, but a recent incident bears her out. When the shuttle *Atlantis* returned from the first Mir docking in early July 1995, it brought Norman Thagard home from his three-month stay aboard the station, along with two Russian Mir cosmonauts, Vladimir Dezhurov and the now-contrite Gennady Strekalov. Russian doctors have a routine for bringing crews home. The first rule is that they should not try to walk, or even stand, when they touch down. Couches are brought up to the returned Soyuz, and the cosmonauts flop into them for a few hours of rest before struggling to stand. But Thagard wanted to get on his feet as soon as *Atlantis* had rolled to a halt. For him, it was a personal challenge, to overcome stiff joints and dizziness and to prove that he could walk despite his long sojourn in weightlessness. His Mir companions thought he was trying to show them up as inferiors.

**TRAINING FOR SOYUZ**
NASA astronaut Shannon Lucid practises for an emergency water splashdown in a Soyuz capsule trainer before her trip to Mir in April 1996. Survival procedures are familiar to her from shuttle training. For any Americans training in Russia, the greatest challenge is not the Soyuz hardware, but learning to speak Russian.

McDonnell's inquisitive anthropologist discovered that Dutch and French members of the ESA Spacelab program had an altogether more pressing problem on their minds. They worried that mealtimes were not taken seriously enough aboard NASA shuttles. They liked good food and treated dinner time as a social break. American fliers seemed to eat just because it was time to refuel their bodies. There are some very basic questions about food on the International Space Station that must be addressed: will everybody eat at the same time? Will the smell of one nation's preferred cuisine wafting through a 'trapped environment' drive other nationals crazy? The official language aboard the space station will be English, but should this requirement be relaxed at meal times?

The Europeans also despaired at NASA's fondness for acronyms, especially as those acronyms tended to mean different things at different times. And anyway, after a while they all started to look the same. SME, SPS, PPS, DPS, TDRS, RCS... If there was an 'S' on the end, that probably meant 'System', but you could never be too sure.

Instrument layouts proved another area rich in lethal ambiguities. British switches 'switch' on and off, while American switches 'turn'. The American custom of 'up' for 'on' is not universally recognized by other nations. This could be critical in space. Flashing red warning lights inform NASA crews that something is wrong. To the Far Eastern mind, red sometimes symbolizes good fortune and its use on instrument panels does not always conform to western standards.

Then there are subtleties of language to do with flight-deck procedures. No matter how accurate the translation of some term might be, its meaning can be radically different. Americans often use the word 'control' for switches, levers and dials: 'thrust controls, attitude controller', and so on. To Russians, this word has only one administrative meaning: 'mission control' or 'state control'. Ask a cosmonaut to locate a steering control and he will think in terms of some authority in Moscow that determines which direction he should fly in.

## HUMAN SPECIMENS

There is one thing that crew members from all nations can agree on: living in space for long periods isn't very good for your health. 'Space medicine', the study of the effects of weightlessness on the human body, has preoccupied scientists since the earliest days of space flight. Continuing research in this area dominates current and future space station science programs. Harvey Willenberg, a scientist for Boeing, sums up one of the challenges for space medicine in the future: 'Can you survive a year in space? Yes, you can. But can you spend a year getting to Mars and be productive when you get there? That's the real question.'

The other issue for space medicine—an expensive practice—is how to return some benefits to taxpayers and commercial interests down on the ground, here and now, instead of worrying about the Martian explorers of some future generation. Medical researchers see astronauts on a space station as flying laboratories in their own right. The weightless environment makes biological systems do strange things, from which a great deal

**LIFE IN SPACE**
This Boeing artwork shows the crew of space station Freedom at work in the science module. In the current design, the modules have been shortened by up to 40 per cent in order to reduce the distances between airlocks at each end. If a micrometeoroid punctures the station, the crew must be able to move immediately to an undamaged module. Debris from old rockets, or even flecks of paint travelling at 13km (8 miles) per second could cause tremendous damage. Apart from this, the painting gives a good idea of what life in space will be like from the year 2002, when the new space station is complete.

about normal functions can be learned. This possibility now constitutes one of the principal scientific justifications for building the International Space Station in the first place. Many of the men and women who wish to stay in it will have to sacrifice their time, their privacy, their entire physical selves to science: constantly giving blood, wearing sensors, logging their food and drink, storing their waste for study and submitting themselves and each other to all kinds of indignities, and perhaps some dangers. During the first shuttle docking mission with Mir, astronaut Bonnie Dunbar suffered a brief but alarming allergic reaction to an injection she had been given prior to the flight in readiness for a weightless medical experiment.

Veterans of the Apollo lunar flights, or the (surviving) early Soyuz test pilots might hanker for simpler days, when space-faring was more romantic. NASA astronaut John Young flew in Gemini, then in two Apollo missions. In 1981, aged over 50, he commanded the first shuttle flight, testing the new ship to see what it could do. In December 1983, Young made what turned out to be his last venture into space. Perhaps it was time for him to yield to younger astronauts, but a telling little incident is worth recounting. It was Spacelab's debut, with ESA's first astronaut, Ulf Merbold, in charge of European experiments. Young put the shuttle *Columbia* into a perfect orbit, and then spent most of the mission on the flight deck. Back in the cargo bay, Spacelab's scientists asked Houston if they could perhaps tell Young to come and lend a hand? Houston asked Young, how about it? With a contemptuous drawl, this pilot, veteran of countless adventures in supersonic jet planes and then in space, a man who had walked on the surface of the moon, replied, 'Hey, I *fly*. I don't do science.'

Inhabitants of the International Space Station will not be able to afford such romantic sentiments.

## THE DRAWBACKS OF WEIGHTLESSNESS

The most serious medical problems caused by weightlessness occur within the structure of the skeleton. On earth the constant shock of walking, running or just standing still seems to stimulate bones into replenishing themselves. In space, bones shed calcium at an alarming rate, and crews are afflicted by something similar to osteoporosis, a condition usually associated with growing old. Bones become brittle. In the long-term, nobody is sure whether or not a crew out in space for, say, three or four years, could ever recover their lost calcium. Their skeletons might become so weak that a return to earth would be impossible. Meanwhile the shed calcium circulates in the bloodstream, and a heavy proportion accumulates elsewhere, for instance as kidney stones.

As well as bones, many muscles atrophy, partly through physiological side-effects but also through sheer lack of use, especially in the lower body. Astronauts call this the 'chicken leg' syndrome. By contrast, arm muscles retain their vigour because astronauts use them for pulling and pushing themselves around. Meanwhile, the cardiovascular system also changes. Blood pools in the head, causing a puffiness around the eyes, headaches, drowsiness and nasal congestion. The body's internal regulation systems misinterpret this as a fluid excess, and excrete as fast as they can. Thirst levels are dampened, and inadvertent dehydration becomes a danger unless astronauts are reminded to drink regularly and often. Blood cell production within the bone marrow is reduced, leading to anaemia and a weakening of the immune system. The heart atrophies and becomes used to a much-reduced pumping load, which can be very dangerous when a return to earth pushes the load up again.

The sense of balance also goes haywire. Intense nausea is common among even the hardiest space crews, at least in the first few hours of weightlessness. Incompatibilities between the inner ear's balance and the visual senses can cause strange illusions.

But there are some indications that the body eventually adapts, finding a new balance in the unusual environment of space (along with the senses). Crews who have been in space for a week or so are often much weaker on their return than those who have been up for several weeks. What's more, astronauts have long made use of certain exercise regimes to simulate the bone shock required to keep the skeleton healthy and to give their hearts plenty of work. Certainly, cosmonauts who have flown for a year or more find returning to earth uncomfortable, but they recover their usual state of health in a matter of weeks, as far as we know. What are the long-term effects of spaceflight? These have yet to be determined, but there is evidence that a slight overall reduction in health can result.

## BAD FOR ASTRONAUTS, GOOD FOR US

The bone disorders that afflict astronauts in weightlessness may well justify the expense of sending them into space in the first place. Nearly 30 million Americans suffer from osteoporosis. The total costs of medical treatment associated with this condition amount to perhaps $60 billion a year: more than twice the space station's overall development budget. If weightless astronauts can be studied, and further secrets of bone chemistry are revealed, then the station may eventually pay for itself by contributing significantly to medical knowledge.

Osteoporosis usually develops over some years. In space, parallel symptoms appear to be greatly accelerated. As early as the second day in space an astronaut's bloodstream will show high levels of calcium shed from the bones. But the skeleton is a dynamic structure, constantly replenishing itself. New calcium from food intake is collected and harnessed in the bone matrix. It seems that the replenishment of calcium tries—and fails—to keep pace with the rate of loss. On earth, the timescales involved in these processes make rapid discoveries difficult. In space, an astronaut's exercise regime, diet, or any number of other factors can be artificially altered, and rapid short-term changes in the bone processes are measurable over days and weeks rather than years. In fact, all the side-effects induced by weightlessness can be exploited to advance learning about the human body, by comparing its behaviour in space to its performance on earth.

PROTEIN CRYSTAL PROCESSING
20.001 °c.
27.054 °c.

# THE PAYLOAD

*Throughout the space station's development, the greatest problem has been defining its ultimate purpose and usefulness. Those who once dreamed of establishing a foothold for deeper exploration of space have sacrificed many of their ambitions for a more inward-looking program, at least for the near future. Apart from political considerations, medical experiments have emerged as the main reason for building the station.*

**EXPENSIVE LABORATORY**
This is the interior of the American Laboratory Module. The four circular panels cover crystallography experiments, and in the foreground a glove box facilitates the safe handling of delicate objects or biological samples.

# THE PAYLOAD

Putting scientist-astronauts into an unusual environment will provide a valuable opportunity for study across the entire range of human biology. But the people who will fly aboard the forthcoming space station will need to be very healthy if they are to withstand the detrimental effects of weightlessness and carry out their duties effectively. Gradual and perhaps reversible conditions like osteoporosis can be studied with their help, but medical researchers will also wish to examine fundamentally unhealthy tissues in space: in particular, clusters of cancer cells. For this, fit space crews will be of little use. Samples from other sources will have to be taken into orbit and studied.

## ROGUE CELLS

Cancer cells are grown as a matter of course in biology labs. There are two principal methods. Cancers can be induced in living organisms (mice, rats and other animals) and then studied, or else, cancerous cells can be isolated and grown in culture dishes. Both methods have their advantages and disadvantages.

Quite apart from the increasing weight of moral complexity attached to vivisection, deliberately inducing cancer in an animal means that the growth processes cannot easily be studied while they happen. The animal has to be dissected after the cancer has started to run its course. By contrast, growing isolated cancer cells in a dish of nutrients enables scientists to examine the cells minutely while the cancer takes hold. But here, the problem is that a dish provides a flat layer of cells, whereas in natural body tissues cancer tends to clump in three-dimensional tumours. The dish-grown cells behave in a different way from 'real' cancers, and are therefore not so valuable for study. Vats of nutrients provide less flattened growths, but the effect is still far from satisfactory.

In space, cancer cells can be grown as solid clumps, not in dishes, but in transparent nutrient-filled canisters. Then they can be studied three-dimensionally while they grow. There will be no gravity to flatten them, and they can be artificially stimulated to behave like growths occurring deep in the body. Treatments, from radiation to lasers and chemotherapy, can then be tried out on the samples.

## MOLECULAR BIOLOGY

Biochemists think in terms of 'formulae' specifying the numbers and ratios of separate chemical elements in a molecule, but this information does not always constitute a thorough description. Apparently insignificant differences between 'left-handed' and 'right-handed' molecules can be of great importance. For example, L-Dopa, a drug compound used to alleviate the symptoms of Parkinson's Disease, is completely diferent in its effects from R-Dopa, the same chemical with a right-hand twist in its assembly. Conventional analyses of such substances often involve breaking them down, which means losing much of the structural information.

Proteins, the most prolific and important substances in the body, can easily be broken down and analysed to reveal the exact numbers of chemical elements that create them. But these elements, carbon, oxygen, hydrogen and so on, are just the building blocks. Scientists will wish to determine the precise structural relationships within individual protein types. Many proteins can be so complex that the problem of analysing them goes far beyond simply determining any left–right bias. DNA is the exemplar of such complexity and also the central target of medical research in space.

In 1953, when Francis Crick and James Watson deduced the double-helix structure of DNA, its constituent chemicals were well known, but few people understood what *structure* they created and how this structure functioned during DNA replication. Crick and Watson took barely 18 months to work it out, revealing the now-familiar double-helix pattern; but they relied heavily on the work of researchers who had already performed laborious experiments with X-Ray crystallography.

Proteins are normally delicate and subtle structures, but under certain circumstances they can be grown as solid crystals, 'freezing' their assemblies in a form durable enough to withstand investigation. DNA can be studied in a crystallized form. In the early 1950s, X-rays punching through DNA-derived crystals revealed fascinating hints about the structure, but only in the form of smudgy interference patterns recorded on photographic plates, all but incomprehensible to the untrained observer. Nevertheless, these crude X-ray experiments provided enough structural data to put Crick and Watson on the right track. What kind of structure might produce the observed interference patterns and be consistent with the known chemical elements within DNA?

Modern electron microscopes and other instruments can now produce computerized visual interpretations of molecular arrangements, taking much of the guesswork out of crystallography. The only real problem remaining to analysts is obtaining pure, tidy crystals to study. In space, free from the distortions of gravity, protein crystals can be grown that are far crisper in definition and geometric precision than equivalent samples grown on earth. Analysis of their structures will become vastly more reliable and useful than the smudged photographic plates pored over by Crick and Watson. Data from space station experiments will assist ground-based genetic research into hereditary diseases, as well as many cancers and bacteria-derived illnesses.

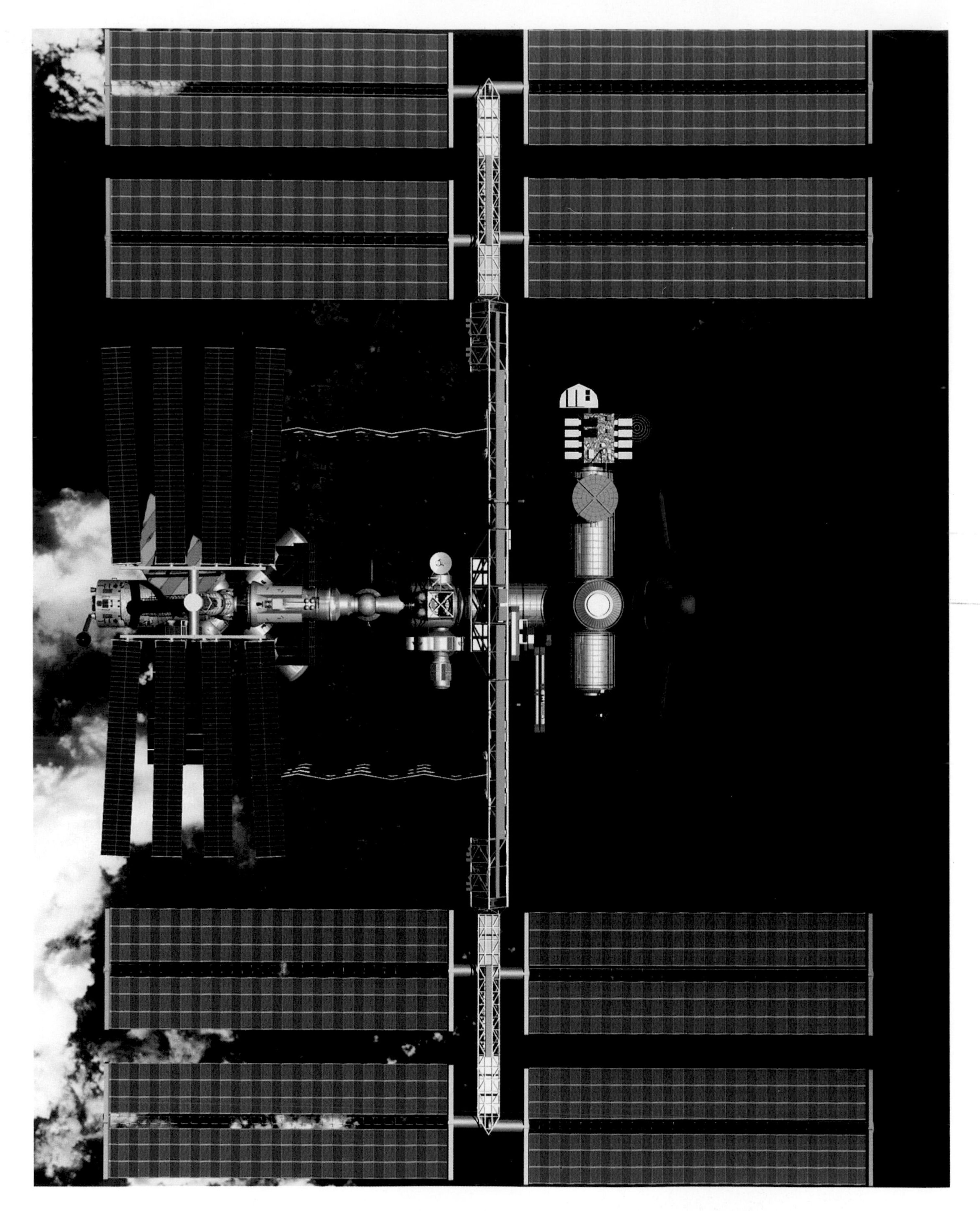

**POWER FOR SCIENCE**
This overhead view of the space station clearly shows how the Photovoltaic Arrays dominate the structure. The main arrays are 74m (243ft) wide, and generate 110 kilowatts between them. The smaller arrays at the right of this view are Russian solar cells, designed to power their Service Module and FGB elements. At the left, a shuttle can be seen docked to the cluster of American, European and Japanese modules.

## FAST CHIPS, FASTER MONEY

The logical extension of growing organic crystals will be to grow inorganic crystals. For instance, if silicon crystals can be manufactured that are even purer than the current products of the electronics industry, then the further miniaturization of computer circuits will be possible. The smaller a circuit is, the faster it will operate, because the substances of the machinery will not impede the electronic currents more than absolutely necessary. Even at speeds close to that of light, electronic pulses travelling through circuitry take some slight time to travel. Reduce the distances, and that speeds up the system. Speed, memory capacity and reliability: these are the Holy Trinity of the electronics industry, and designers are seeking an inevitable ideal. They would like to make computers whose smallest logic elements are single atoms. In theory this has been proven possible, but only if tiny electronic circuits can be fabricated within utterly pure, utterly regular crystals.

Small quantities of extremely specialized chips for research purposes may soon generate a 'low-volume, high-turnover' industry in space. A shuttle flight could easily accommodate a small box containing a few hundred chips, each worth many thousands of dollars. But large-scale orbital manufacturing of such ultra-pure chips in their hundreds of thousands for conventional computers is unlikely for the time being, not because they cannot be made cheaply in space, but because they cannot be transported cheaply back down to earth. The same will apply to biotechnology products. Small phials of exotic samples may be worth a thousand times their weight in gold, not as consumer drugs in themselves, but as fundamental drivers of research that will aid the development of new drugs on earth.

The principal product of a space station, profitable or not, will be information, which can be beamed back to earth at little cost. It is worth keeping in mind that most notions of huge orbiting industrial complexes fall down when it comes to factoring in the transportation costs of raw materials and physical products. The development of a genuinely cheap space transportation system will be essential to any proper space industrialization process. In the meantime, and despite the current high costs of space shuttle launches, weightlessness should provide a very productive and potentially profitable research environment.

Except there is no such thing as weightlessness.

## THE MYTH OF ZERO-G

A spaceship in orbit around earth is not free of gravity. Far from it. Gravity is what keeps it from flying off into the depths of space. In strictest terms, the ship and its crew are in free-fall like a lift carriage filled with passengers just after the cables have snapped. The station falls endlessly around the planet, its tendency to fly off exactly balanced by the earth's tendency to pull it down. In low orbit, a space station's principal problem is that slight atmospheric drag gradually allows gravity to win. This was why the early Salyuts kept falling to earth after relatively short stays in orbit, until Russia learned how to reboost them. The American Skylab stayed aloft for several years because it was originally placed into a higher orbit than the Salyuts, but after six years it still succumbed to drag. The new space station will require reboosting at 90-day intervals in order to stay aloft for 12–15 years.

Atmospheric friction apart, the net balance between the station's centrifugal tendency to fly outwards and gravity's attempt to pull it inwards cancels everything out as far as the occupants are concerned. But a gravitational effect is still measurable. For instance, the parts of a station closest to earth are under a greater gravitational pull than the areas away from earth, just as Isaac Newton tells us. In addition, the station has its own mass: 462 tons, to be precise. This is enough for it to generate a tiny gravitational field of its own. As with any celestial body, the usual rules apply: the station's own gravity is strongest in the center and weakest at its outer extremities.

Space researchers are accustomed to thinking in terms of 'micro-gravity' rather than 'weightlessness'. The precise location of experiments on the space station thus becomes important. Another difficulty is that the station's mass configuration changes every time the crew moves from one module to another. This shift is very tiny, but measurable even so. A more discernible problem is the slight accelerations in any number of different directions generated by the crew pushing and pulling themselves around, causing the station to react against them. These accelerations are easily measurable, and substantial enough to present a problem. If a micro-gravity experiment is subjected to acceleration, then it experiences, albeit briefly, a force that is absolutely indistinguishable from gravity.

Imagine you are on a little rowboat, in the middle of a calm lake. For mysterious purposes known only to her, your friend is attempting to keep a glass of water that is full to the brim from spilling, so that she can study the surface of the fluid. Unwisely, you clamber over to her side of the boat. This action produces a reaction in the boat, which moves beneath your feet and shakes your friend's glass. So you decide to keep out of harm's way by going for a swim. As you slip over the side, and no matter how much care you take, the boat will always respond to your pushing off against it. Your friend's delicate experiment will be ruined.

NASA planners have devoted a great deal of time and energy to solving the equivalent problems in space. How can the station compensate for astronauts going 'over the side' on space walks? How can experiments be protected against the crews' movements? What happens to micro-gravity experiments when an 80-ton shuttle docks to the station and shifts the collective center of mass, as well as juddering the whole structure when the docking latches clunk into place?

The primary solution will be to reserve 180 days out of every year for microgravity research. On these days, often scheduled in weeks-long stretches, shuttle dockings, new module arrivals and so forth will be banned, and crews not engaged on experiments will be kept occupied on other work that keeps them reasonably still. During the other 180 days, shuttles, Progress ferries and new modules can dock, astronauts can clamber around the walls, and vigorous exercise-related medical experiments can be conducted.

**THE 'POWER TOWER'**
When NASA began serious space station studies in the early 1980s this design was used as a basis for requests to contractors for more detailed proposals. NASA wanted the main bulk of modules and equipment to rest at one end of the structure, with the less massive solar sails, antennae and sensors at the opposite end. Earth's gravity would gradually pull the heavy end down towards the planet, and a stable position would thus be maintained, with earth observation instruments at the bottom and astronomy payloads at the top, pointing out at space. The design fell out of favour because micro-gravity researchers wanted the laboratory modules to be as near as possible to the station's center of mass.

So far, this allocation of microgravity time is largely a theoretical management problem. How it will work in practice remains to be seen. Even a crew under 'silent running' cannot be expected to stay totally still for days, or even hours, at a time. Microgravity experiments will therefore be housed in special racks, isolated from the space station's structure by springs, shock absorbers and inertia-dampening mechanisms. This approach is not far removed from the vibration-proof platforms of certain very expensive stereo turntable systems available here on earth. Experiments requiring extreme degrees of isolation will be put on special pallet racks outside the main body of the station, held in place by compact shock-absorbing hinges.

But one module will incorporate artificial gravity—though not for the crew's comfort. Small biological samples, plants, micro-organisms and insects, will be rotated in a centrifuge so that the effects of varying gravitational forces on their growth and behaviour can be studied. Every organism on earth is influenced by gravity. Flowers and trees 'know' to grow upwards and send roots 'downwards'. This behaviour is partly reliant on chemical and light-sensitive stimuli, but gravity is a significant stimulus. Gravity's role in the balance perception of insects and spiders can also be studied by subjecting these creatures to different simulated gravities in the centrifuge. In miniature, they will experience the same kind of rotation-induced artificial gravity envisaged by Hermann Noordung and Wernher von Braun when they conceived their giant wheel-shaped space stations.

With so many elaborate experiments packages operating on the space station at any one time, safety will become a major issue. Biological experiments in the centrifuge and elsewhere will have to be carefully restrained in airtight boxes; electrical supplies to all the hardware will need monitoring; and fire will be a significant hazard, particularly as the station's crew will spend some of their time deliberately setting things alight.

## FIRE!

On earth, a candle flame tapers upwards, drawing in fresh oxygen at its base and expelling hot burnt fumes at its tip by convection. In space, a flame is spheroid, and can sputter out in a fog of its own fumes. Fire is still a potential danger, however. Shuttle electrical systems have occasionally produced sparks and fumes, and the solution has always been to switch them off and use back-up equipment. Russian space stations have experienced two serious fires, including one underneath a flooring panel where it could not easily be extinguished. Toxic fumes, rather than flames, pose the principal danger. The question arises: how does a fire-detection system operate in a space station?

Obviously, ceiling-mounted smoke detectors are useless because there is no convection to carry hot combustion products up towards them. Infrared detectors might make a fair substitute, but what if a fire has broken out in a hidden recess, or in an enclosed box of electrical circuits? Heat sensors will be placed all around the station's hardware, but again, with no convection to aid them they could easily miss a fire right in front of their noses, so to speak. The study of how flames behave in space will be crucial for the future design of spacecraft, and even for safety adaptations to the station itself.

Meanwhile, on earth, 85 per cent of all our energy production involves some kind of burning: petrol, coal, gas, wood. As the vast tonnage of carbon-based combustion products build up in the atmosphere, contributing to the so-called 'greenhouse' effect, so the search intensifies to make these burning processes more efficient and less toxic. If typical combustion systems could be improved by just 2 per cent in efficiency, the American economy could save $8 billion a year in fuel bills. The study of flames in space will provide scientists with data about how combustion works at its most fundamental level. In fact, substances don't actually have to burst into flames to elicit the scientists' interest. They just have to change temperature sufficiently to melt, to vaporize—or to freeze.

**CORE OF THE STATION**
A close view of the central cluster of pressurized space station modules: at left, US, European and Japanese logos can be seen on their respective modules. The Russian Service Module is at bottom right. The entire structure seems flimsy, but in weightlessness this is not a problem. However, the docking of shuttles will send vibrations rippling through the whole assembly, and this will impact severely on microgravity experiments. Notice the rectangular array of experiments attached to the Japanese experiment module, isolated as far as possible from the main station structure by thin suppporting arms.

## ALTERED STATES

When ice turns to liquid, and liquid turns to gas, these three familiar states are quite distinct from each other. But the precise transition from one state to another creates a fleeting, little-understood state of matter completely different from our conventional experience. This borderline condition is called 'phase transition'. Convection makes detailed and reliable study of phase transition difficult on earth. In space, this delicate area of atomic physics can be studied in an ideal environment in which no forces other than the phase transition itself are at work.

Physicists are keen to explore phenomena such as phase transition so as to expand the frontiers of 'pure' (as opposed to 'applied') scientific knowledge. Geneticists look forward in more pragmatic fashion to applying station research to the health care and biotechnology industries on earth. Engineers will study the dynamic behaviour of the station itself to learn about resonance phenomena, shock absorption, heat transfer and

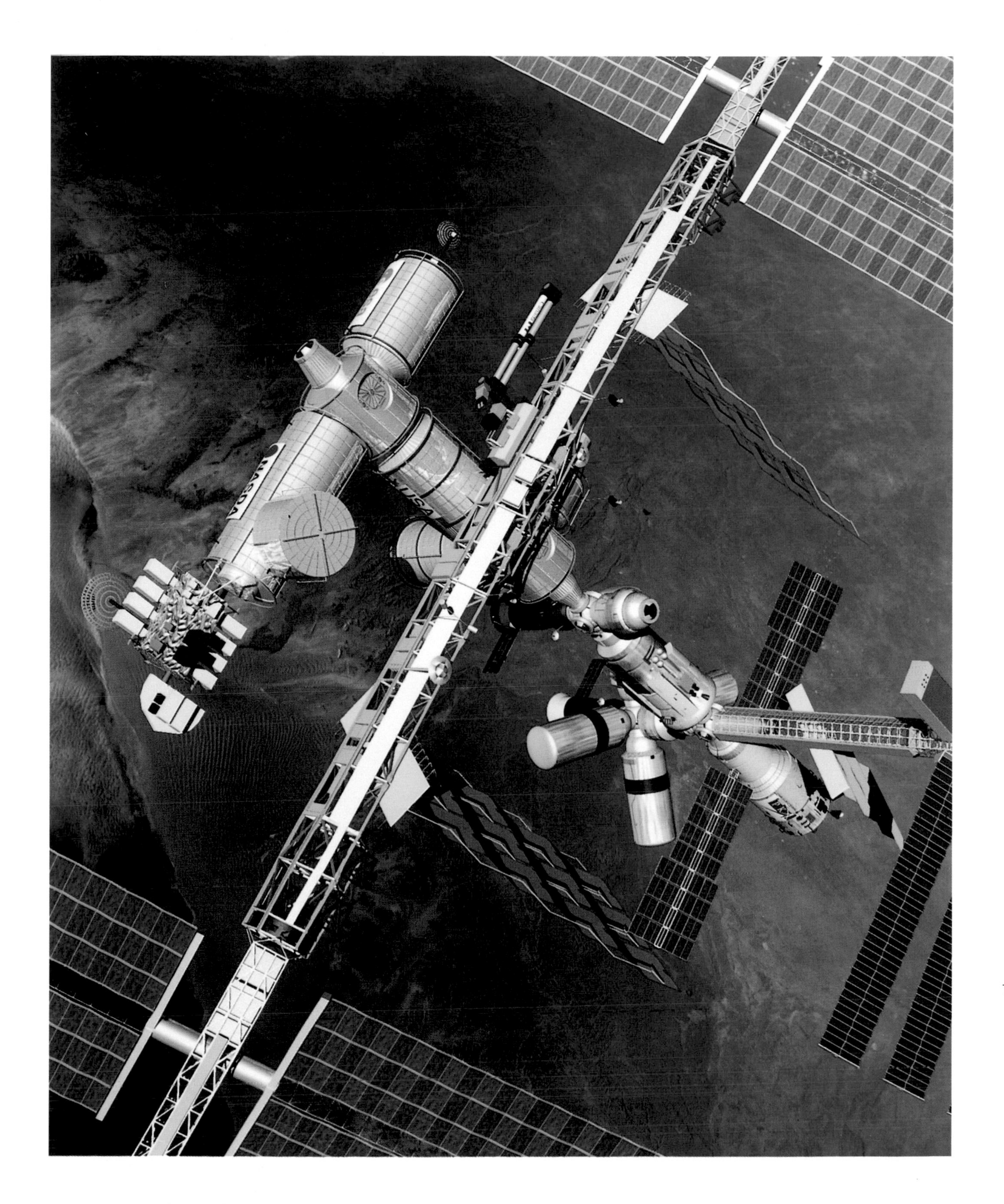
NASDA
ESA

## BIOTECHNOLOGY

| | |
|---|---|
| **Tissue Culture Studies** | Knowledge of normal and cancerous mammalian tissue growth and development processes. |
| **Protein Crystal Growth** | Provides information needed to understand and modify the behaviour of proteins; HIV research. |
| **Separation Sciences** | Separation and purification of biological cells and proteins for disease treatment research. |
| **Cell Fusion** | Genetic engineering of agricultural plants to improve yield and increase resistance to disease. |
| **Collagen Processing** | Generating materials for use in repairing wounds, or other treatments for human connective tissues. |
| **Isoenzyme Production** | Shortening production times for manufacturing proteins used in human disease treatment. |

## LIFE & BIOMEDICAL SCIENCES

| | |
|---|---|
| **Gravitational Biology** | Studying the role of gravity in all life on earth, from bacteria to plants, animals and humans. |
| **Space Physiology** | Knowledge of human balance and locomotion functions and perception; immune responses to microgravity. |
| **Closed Life-Support** | Recycling of waste and respiration products; air, water and consumables replenishment. |
| **Operational Medicine** | Study of cardiovascular performance, muscular and vestibular disorders; osteoporosis. |
| **Human Behaviour** | Procedures to help with sleep dysfunction; modelling of human performance; team psychology. |
| **Crew Health Care** | Biomedical monitoring; remote-care technology; compact health care systems for emergencies. |

## EARTH & ATMOSPHERIC SCIENCES

| | |
|---|---|
| **Oceanic Research** | Monitoring sea surface temperature, wind speed, sea roughness, ocean currents, ice coverage, etc. |
| **Atmospheric Research** | Analysis of gases and aerosols; measurement of clouds, ozone and trace gases; storm monitoring. |
| **Near-Earth Studies** | Measurement of global radiation exposure from gamma-ray bursts and solar particles, etc. |
| **Resource Monitoring** | Trial techniques for monitoring crop health, plant distribution, animal populations; plankton density, etc. |
| **Transport Patterns** | Possible analysis of the impact of global air and sea traffic; aircraft exhaust trails; ship wakes. |

## ENGINEERING & TECHNOLOGY

| | |
|---|---|
| **Human Support** | Designs for firefighting suits, diving equipment, power tools, and special garment cooling. |
| **Structures Research** | Modelling and analysis, vibration damping for buildings, bridges, airplanes and other structures. |
| **Spacecraft Materials** | Lightweight oxygen tanks; anti-corrosion pipes; self-healing paints; solar cells; mechanical component lubrication. |
| **Communications** | Commercial requirements for new technology to maintain a lead in space-based communications. |
| **Information Systems** | Protection of electronics against radiation, for defence and civil aerospace applications. |
| **Fluid Management** | Flow phenomena; long-term storage; cryogenics; surface tension and capillary phenomena. |

## MATERIALS RESEARCH

| | |
|---|---|
| **Casting Processes** | Producing defect-free castings for precision engineering components, particularly rotary elements and ball bearings. |
| **Foamed Metal** | Lightweight high-strength aluminum and other metals for aerospace industry applications. |
| **New Alloys** | Metals in new alloy combinations unavailable on earth due to convection effects during casting. |

## POLYMERS & CHEMISTRY

| | |
|---|---|
| **Encapsulation** | New technology for timed-release drugs dosage and implementation of hormone treatments. |
| **Diffusion** | Studying the fundamental physics of diffusion free from the effects of convection phenomena. |
| **Polymerization** | Research into the 'weak' atomic forces involved in polymerization processes. |

## FLUID PHYSICS

| | |
|---|---|
| **Containerless Working** | Investigating procedures to handle fluid and transitional materials free from a container's walls. |
| **Heat Transfer** | Studying heat-transfer phenomena within fluids unaffected by convection. |
| **Viscosity Studies** | Examining the chemical and engineering implications of delicate surface tension phenomena. |
| **Phase Transition** | Investigation of non-combustion fluid phase-transition phenomena; freezing; evaporation; condensation. |

load distributions, all of which can be applied to conventional aerospace applications. Looking further ahead, NASA and the other major space agencies will formulate databases that will help make the next generation of spacecraft yet more efficient. The benefits of space station research promise to be wide and varied, but will they be worth $20 billon? That all depends on the definition of 'worth'.

## WHY DO IT?

Private industry has been slow in picking up the bill for creating a laboratory in space. This is why taxpayers' money from all around the world has had to take the lead, through the various government space agencies. With the space station now advanced, and scientific payloads under development, countless institutions are now eager to become involved in research projects. In America more than 600 scientists are co-ordinating experiment schedules with 35 major industrial research companies, and over 80 universities and educational establishments. Detailed proposals for 1,800 experiments have already been evaluated.

**BRINGING HOME THE LOAD**
Shuttle flights are complex and expensive. Cheaper, simpler means of transferring loads to and from the space station will have to be developed. Transportation costs are the main barrier to profitability in space. This speculative artwork shows an Orbital Transfer Vehicle (OTV) which would be capable of bringing payloads home in a shielded reentry pod. NASA and ESA both wish to develop OTVs with manned and unmanned capability.

Bearing in mind that—according to its detractors—nobody apart from NASA ever really wanted a space station in the first place, a great many people seem more than happy to take an interest now the project is under way. Even so, respectable scientists from around the world have spent much of their time criticizing the space station's expense and suggesting that better experimental work can be done more cheaply by robot satellites, or even down on the ground. Are they right? The best argument in the station's favour is that \$20 billion isn't that much money in the great scheme of things. For instance, America's defence budget for 1995 alone was \$252 billion.

Another way of looking at it is this: a station will cost each taxpayer about \$9 a year. Even so, \$20 billion cannot simply be spent for the fun of it. Jobs, politics, international diplomatic relations, and the prospects for medical research are all critical pieces in the space station jigsaw, but none of these factors properly explains, more than a decade after the project was initiated, *why* we decided to build it in the first place.

As the scientific users of the station have won political influence over NASA's space cadets, the potential microgravity disruptions of heavy-duty activities like assembling moon ferries and Martian ships alongside the station have become unwelcome. In practical terms, this is inevitable and probably appropriate. The space station will be expected to pay its way, delivering tangible benefits to the taxpayers who are supporting its costs. But all the scientific and commercial justifications that have emerged since former president Reagan's 1984 announcement are just that: justifications for a project originally founded on abstract instincts rather than strict logical requirements.

The same triumph of emotion over reason applies to the entire history of manned space flight, from the theorists who showed that it was possible, to the politicians who authorized the expenditure. In 1969 the novelist and journalist Norman Mailer investigated the Apollo moon project, trying to determine its meaning. In *A Fire on the Moon*, he concluded, 'For the first time in history, a massive technological bureaucracy had committed itself to a surrealist adventure, which is to say that the meaning of the proposed act was palpable to everyone, yet nobody could explain its logic.'

Of course a big space project will always excite the forward-looking imagination. In broad instinctive terms a continued expansion into space *seems* like a good thing thing for humanity to achieve. But we are left with the remarkable fact that the forthcoming space station is the most complex and expensive international collaboration ever attempted in peacetime history, and yet its immediate scientific, political, economic and strategic benefits remain vague. The station may not be able to 'pay its way' in any practical financial definition of that phrase. For once, we seem to be promoting a sense of collective cosmic destiny ahead of our more earthly concerns. Perhaps the best reason for building the station is simply to prove that we *can*, and that so many nations can work in harmony on a common project. If the station provides a truly successful example of global collaboration to inspire future generations, then \$20,000,000,000 will have been very profitably spent.

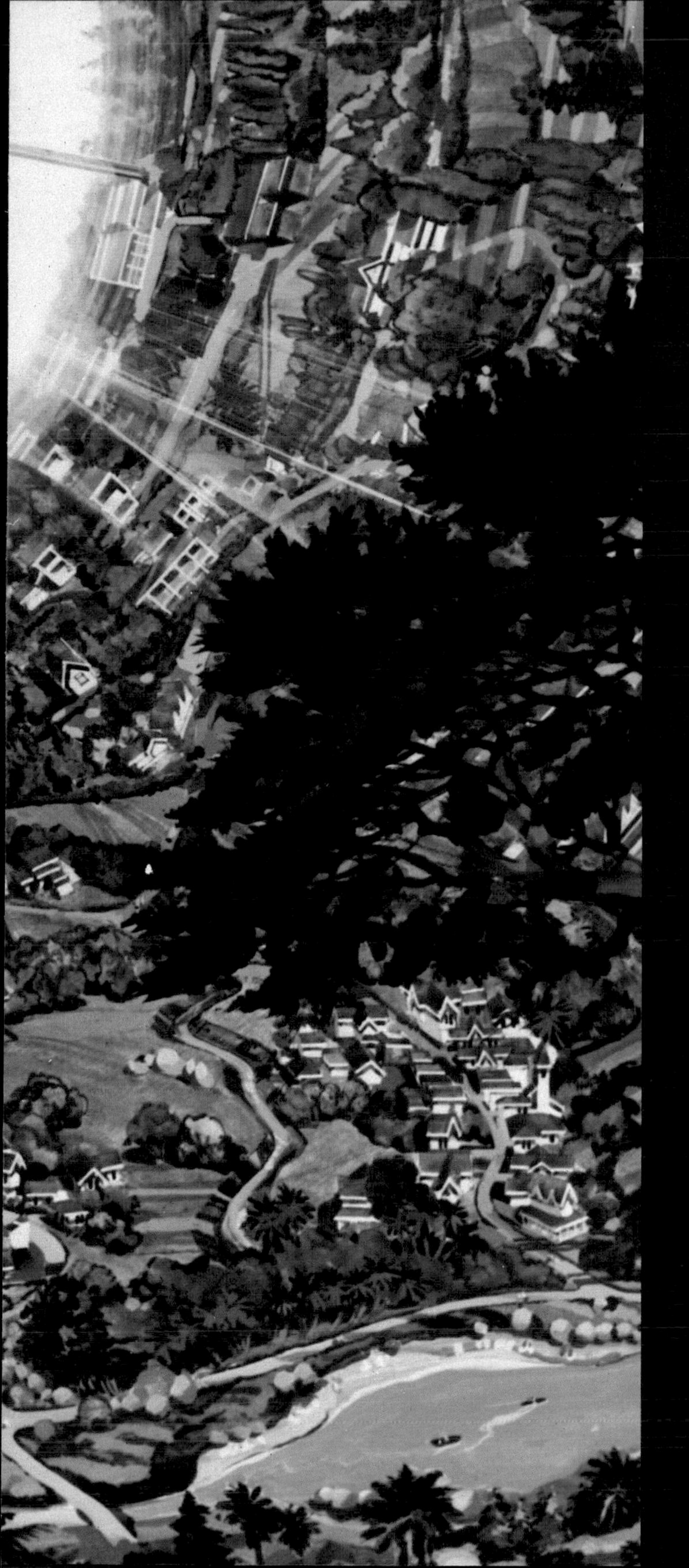

# PAST AND FUTURE

*The current space station owes its origins to ideas dating from long before the first rockets were built, let alone sent into orbit. On the following pages some key ideas are represented. And what of the future? Will we build even larger stations one day? Will we generate power from the sun, build orbiting hotels, or perhaps establish artificial new worlds to live on?*

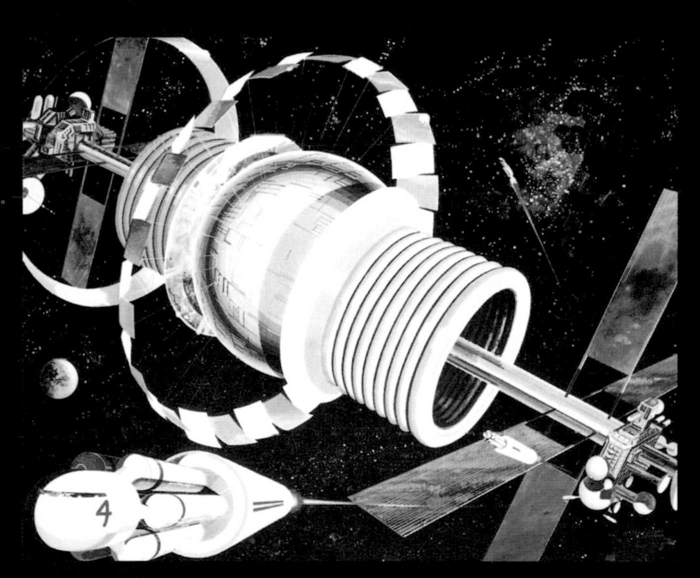

**A WORLD IN SPACE**
This gigantic rotating cylinder accommodates 10,000 people. They are not on a tour of duty. They live here permanently. This concept is derived from the work of Gerard K. O'Neill in the 1970s.

# PAST AND FUTURE

**NOORDUNG'S WHEEL**
The first diagram of a space station ever to be published, as it appeared in Hermann Noordung's *The Problem of Space Travel* (1927). There is a lift-shaft, backed up by two curving stairways. The whole structure is supported against centrifugal stresses by spokes, just as on a bicycle wheel.

The first description of a 'space station' probably owes its origins, in literature at least, to the American author Edward Everett Hale. In 1869 he wrote a story called *The Brick Moon*, in which he describes a brick sphere, 60m (200ft) in diameter, catapulted into space between two giant rotating flywheels. It is intended as a beacon, a reflective blip in the sky that can be observed by ocean-going ships and used as a reference for navigation and timekeeping. Some of the sphere's builders accidentally become passengers when it is hurled into space. They communicate their plight using mirror semaphore. Back on earth, the flywheels are powered up again to send food and supplies up to the stranded workers.

## MAKING THE WHEELS GO ROUND

By the 1920s, astronautical theory was advancing at a great pace. Rocket pioneers such as Robert Goddard in America, or Hermann Oberth and Willy Ley in Germany understood the potential for space stations, but the first thoroughly detailed engineering proposal appeared in Hermann Noordung's book *The Problem of Space Travel* (1927).

Very little of Noordung's background is known today, but his space station proposal survives in all its prescient detail. He described a *Wohnrad* ('Living Wheel') whose gentle rotation would provide its crew with artificial gravity. Airlocks and safety bulkheads were all taken into consideration, along with a huge parabolic dish that would collect and focus sunlight for power. The docking airlock rotated at the same rate as the station, but in the opposite direction, thus maintaining its position relative to the horizon. Rocket ships could approach without themselves having to tumble. Noordung's station was intended as an astronomical observatory and a navigation beacon.

Noordung predicted that weightlessness would play havoc with the human balance system, as well as providing an ideal environment for new scientific research. His design influenced a generation of astronautics engineers, including Wernher von Braun, whose wheel-shaped space station as depicted in *Collier's* in 1952 was remarkably similar in its concept and overall appearance.

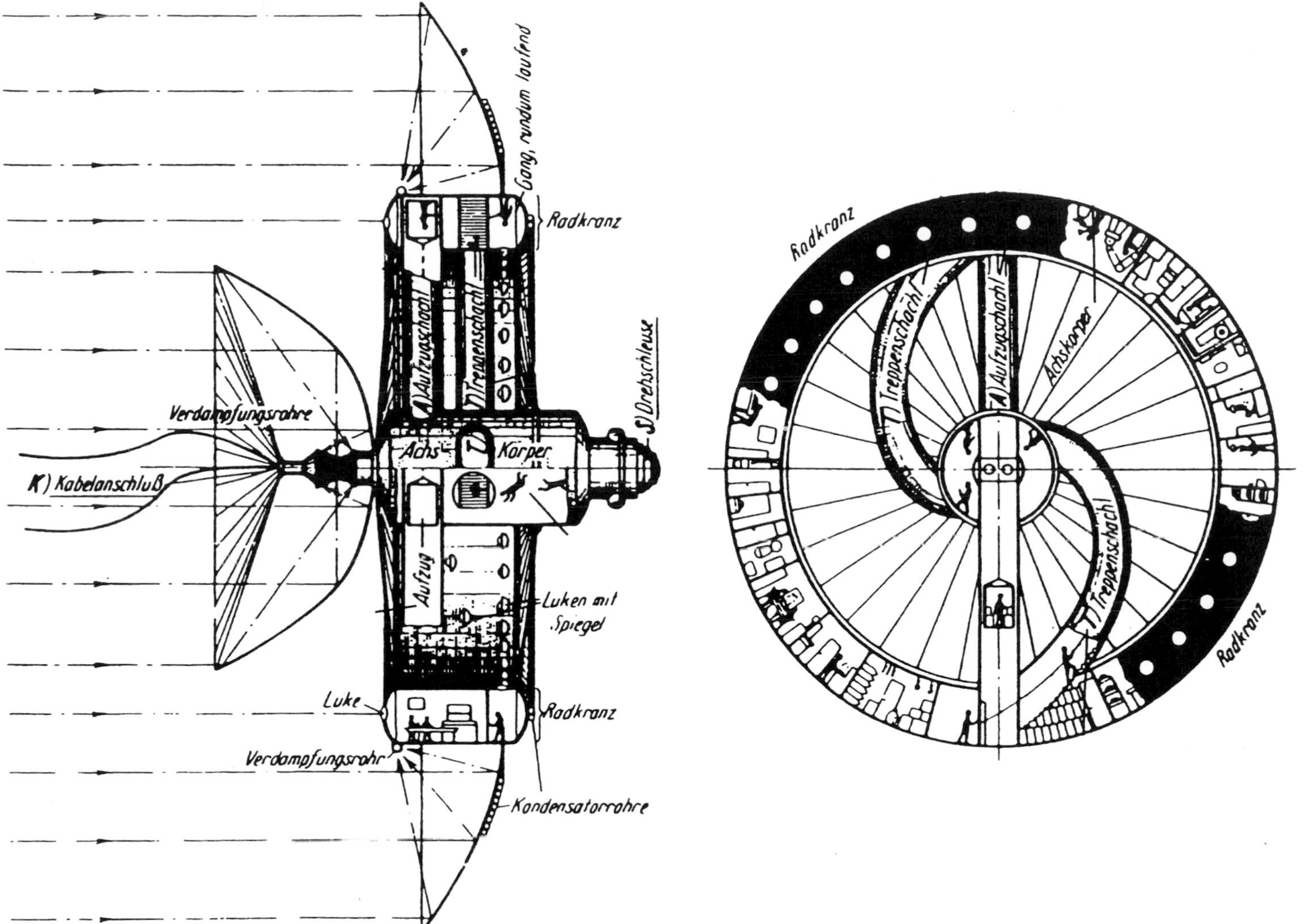


## MAKING SPACE PAY

In the early 1940s Arthur C. Clarke helped develop radio navigation systems for allied aircraft returning from their bombing raids over Germany. At the end of the war Clarke recognized that his dream of space flight would not be achieved unless governments and industrial investors could be persuaded to see its economic benefits. In October 1945 he published an article in an electronics magazine, *Wireless World*, essentially outlining the modern concept of geostationary communications satellites. The only difference between Clarke's ideas of 1945 and the global network that we see today is that he was thinking in terms of very bulky and unreliable 1940s radio equipment. He suggested that his 'Extraterrestrial Relays' would have to be tended by human engineers. In other words, they would be permanently manned space stations.

Clarke was a leading member of the British Interplanetary Society (BIS). Founded in 1933, this amateur but highly disciplined organization remains one of the most influential study groups in the field of astronautics. In 1939 the BIS showed that available technology could be used

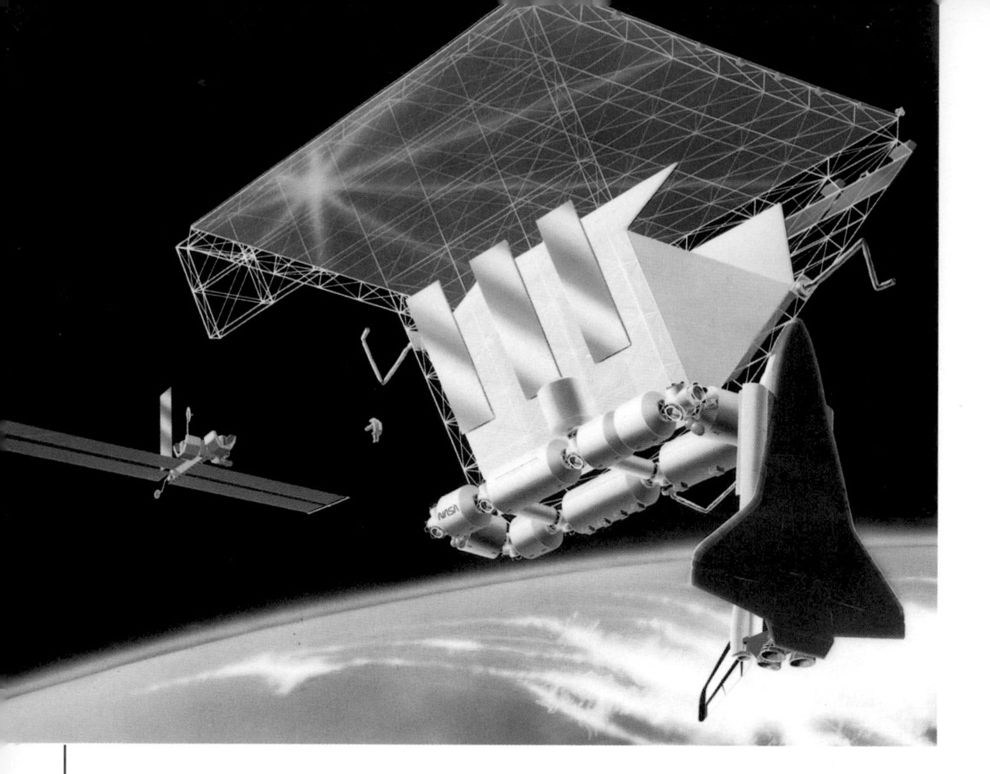

**SMALL-SCALE POWER PLATFORM**
This 1982 idea for a shuttle-serviced space station shows the structure dominated by a large flat area of solar panels, which might be used for experiments in generating electricity from solar radiation. Even a power station this size would be too insignificant to make any difference to our power requirements down on the ground.

to send a rocket to the moon. By 1946 Clarke's colleagues Harry Ross and Ralph Smith had devised a rotating space station similar to Noordung's proposal, a detailed version of the platform envisioned by Clarke as a communications relay, complete with a radio room, a small library, a kitchen, a surgery, cabins for the crew, an office for the Chief Engineer and separate quarters for the Station Director—in many ways, a very English space station.

## POWER FROM THE SUN

Clarke's ideas may have been at least partially superseded by the advent of miniaturized electronics, but another practical use for space stations remains a possibility: gigantic solar power panels could be harnessed for the benefit of people on earth. The collected solar energy could be translated into electricity and used to generate a microwave beam. On earth, microwave dishes gather the radiated energy and reconvert it into usable electricity.

The advantage of this idea is that the sun's power is free, and limitless. The technology for a space power station already exists. The disadvantages, however, are daunting. In order to make any truly significant contributions to earth's electrical requirements, a space power station would require several square kilometres of solar panels, thus putting into question the benefit of building such a large and expensive structure, as set against the potential energy savings. As for the ground facilities, two theoretical models exist—a small receiving dish scooping up an intensely powerful and very concentrated microwave beam, or else a vast area of dishes picking up a weak and widely-scattered beam.

The first version, with its beam as slender as a tree-trunk, is dangerous. What happens if the beam strays from its target? It would cook anything in its path. The alternative, to spread the microwave harmlessly over a wide area is just as environmentally unwelcome. Hundreds, and perhaps

thousands of square kilometres of the earth's surface would have to be devoted to receiving dishes. But the solar expert's frustration is understandable. It is a pity that we cannot make better use of the sun to produce electricity on the ground.

## COLONIZING SPACE

The phrase 'space station' conjures up an image of a metal cabin more or less the size of a three-storey house, manned by a select group of technicians, or perhaps one of von Braun's or Noordung's 'Living Wheels' a few hundred metres across, with several dozen crew aboard. But some notable theorists have imagined gigantic stations bigger than many towns on earth, serving social rather than scientific ends. Huge populations might live off the earth, or so the theory goes.

In 1903 the great Russian pioneer of astronautics, Konstantin Tsiolkovsky (who worked on detailed ideas about rockets from the 1880s until his death in 1935), proposed a huge habitable cylinder, spinning on its axis and containing a greenhouse with a self-supporting ecological system; and in his novel *The World, the Flesh and the Devil* (1929) J.D. Bernal devised 'Bernal Spheres', self-supporting 'Worldships' capable of housing many thousands of inhabitants.

These colossal concepts were revived in the 1970s by an American academic, Gerard K. O'Neill at Princeton University. In the wake of the Apollo lunar landing of 1969, O'Neill proposed huge colonies in space as a way of easing the population pressures on earth. The structures would be perhaps three kilometres in length, fabricated from materials processed in lunar factories. At first, O'Neill's ideas were intended primarily as a theoretical exercise to stretch the imaginations of his physics students, but the space colony theme hit a nerve during the 1970s with its detailed studies of closed ecological systems—an urgent concern for environmental campaigners. Especially popular was O'Neill's proposal that all heavy industry and energy production should be taken into space, thus saving the earth from the burdens of pollution.

Smaller versions of O'Neill's space world might soon be built as tourist venues. It would require only a few hundred rich holidaymakers per year to make a small orbiting hotel worthwhile—provided that a suitable passenger-carrying launch vehicle was already available and safe to fly. Development of a genuinely reusable and fuel-efficient space transport system remains the last great unconquered space frontier. When flying into space is no more expensive or dangerous than stepping aboard a Boeing 747 for a trip across the Atlantic—then the future starts.



## SOLVING THE WRONG PROBLEM

Centrifugal rotation has been a favourite theme for many space station theorists, from Tsiolkovsky in 1903 to O'Neill 70 years later. The spin would produce an artificial gravity. The current ideal is to exploit rather than counteract the weightless environment in order to conduct valuable scientific procedures that are impossible on earth. Where once the gentle spinning of a space station was seen as a way of making life more or less bearable for its crew, total weightlessness is now seen as the main justification for building one in the first place. This is the ultimate explanation for why Skylab, Mir, and now the International Space Station look nothing like the speculative visions of the early space theorists.

Some of the old dreams do seem to be coming true, however. Those people who are frustrated by the apparently slow development of our space effort since the glory days of Apollo might take comfort from an interesting and simple statistic: today, more people go into space every year than during the entire 1960s, at the supposed 'height' of the Space Age. Although we can expect upsets, and perhaps even some slight slippage in the assembly schedule before 2002, the station will be built, and humans will be in orbit to stay.

The real question must be: do we go further into space?

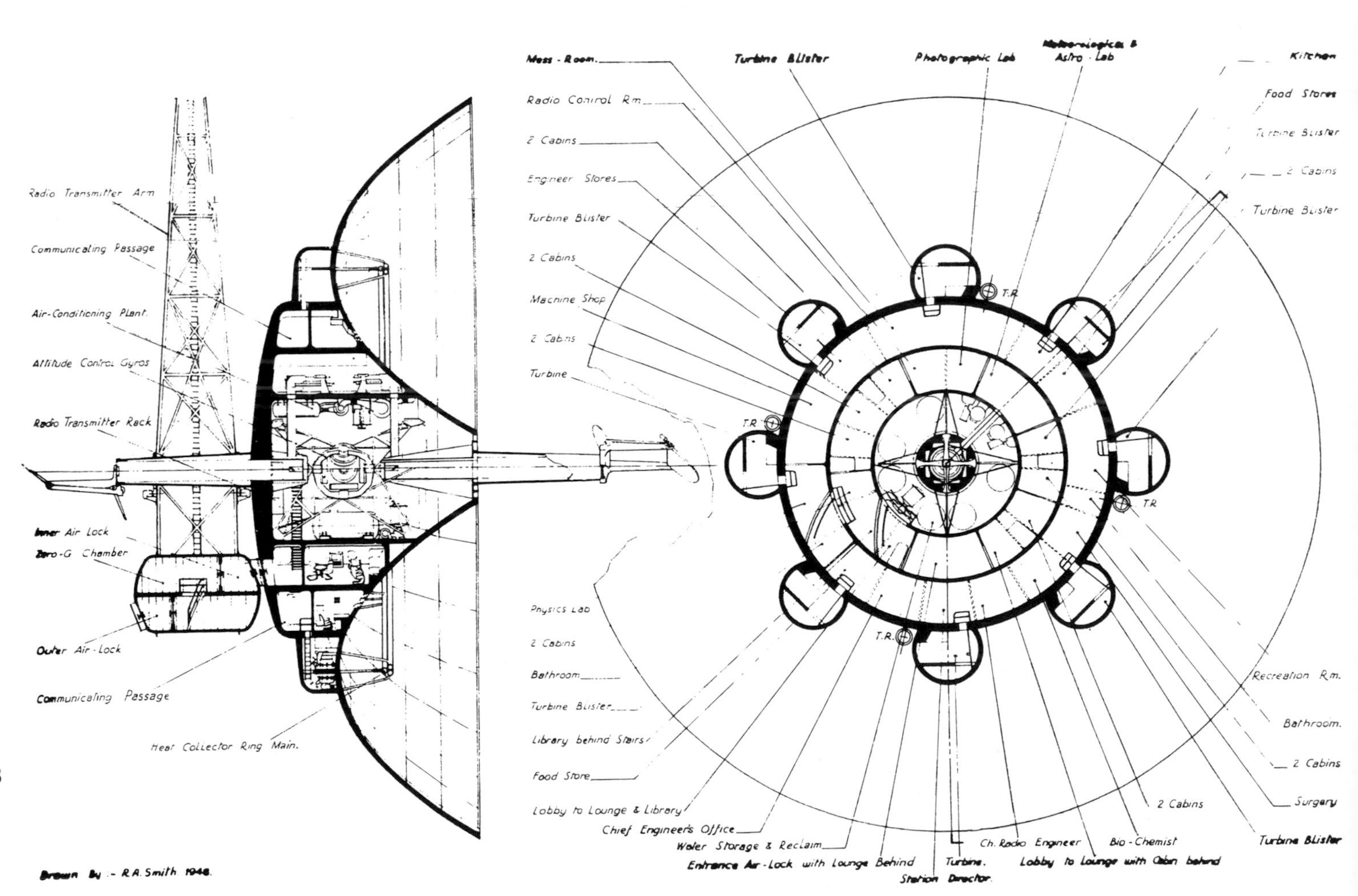


**A BRITISH CONSTRUCTION** Ralph Smith's 1946 drawing of a rotating space station for the British Interplanetary Society is closely detailed.

Atlantis
USA

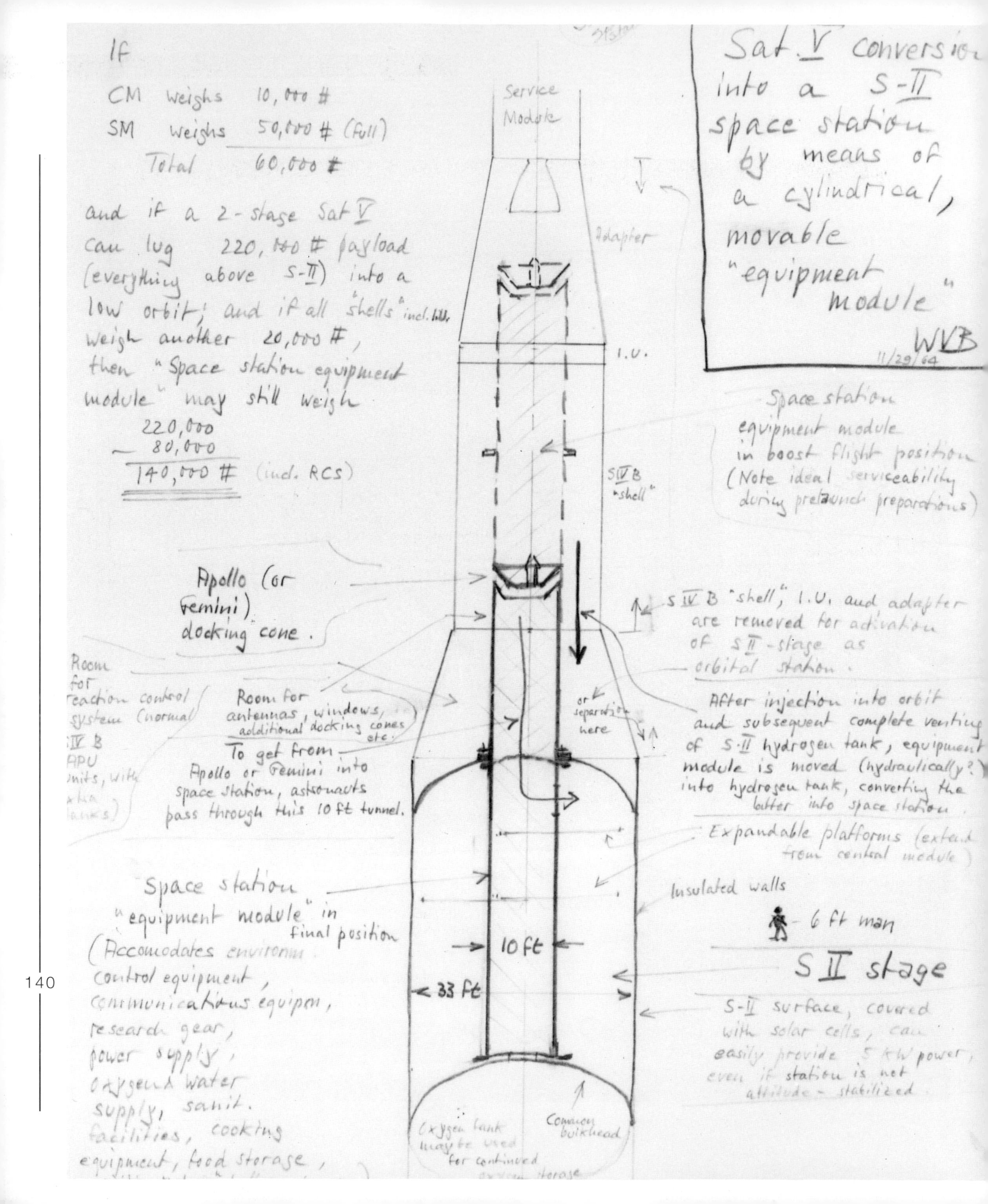
If
CM weighs 10,000 #
SM weighs 50,000 # (full)
Total 60,000 #
and if a 2-stage Sat V can lug 220,000 # payload (everything above S-II) into a low orbit; and if all "shells" incl. bulk. weigh another 20,000 #, then "Space station equipment module" may still weigh
220,000
− 80,000
140,000 # (incl. RCS)
Sat. V conversion into a S-II space station by means of a cylindrical, movable "equipment module"
WvB
11/29/64
Service Module
Adapter
I.U.
SIVB "shell"
Space station equipment module in boost flight position (Note ideal serviceability during prelaunch preparations)
Apollo (or Gemini) docking cone.
S IV B "shell", I.U. and adapter are removed for activation of S II-stage as orbital station.
Room for reaction control system (normal S IV B APU units, with extra tanks)
Room for antennas, windows, additional docking cones etc.
or separation here
To get from Apollo or Gemini into space station, astronauts pass through this 10 ft tunnel.
After injection into orbit and subsequent complete venting of S-II hydrogen tank, equipment module is moved (hydraulically?) into hydrogen tank, converting the latter into space station.
Expandable platforms (extend from central module)
Insulated walls
Space station "equipment module" in final position
6 ft man
10 ft
33 ft
S II stage
(Accomodates environm. control equipment, communications equipm., research gear, power supply, oxygen & water supply, sanit. facilities, cooking equipment, food storage,
S-II surface, covered with solar cells, can easily provide 5 KW power, even if station is not attitude - stabilized
Oxygen tank may be used for continued oxygen storage
Common bulkhead

# SELECTED BIBLIOGRAPHY


**PIONEER PUBLICATIONS**

***The Problem of Space Travel***
Hermann Noordung
Ernst Stuhlinger, J.D. Hunley & Jennifer Garland (editors)
NASA translation, SP–4026, 1995
(Includes the full text of Noordung's 1927 book, and facsimiles of his original diagrams.)

***Across the Space Frontier***
Cornelius Ryan (editor)
New York: Viking, 1952
(Based on the *Collier's* articles, including Bonestell artwork.)

***Conquest of the Moon***
Cornelius Ryan (editor)
New York: Viking, 1953
(Based on the *Collier's* articles, also illustrated.)

***Blueprint for Space***
Frederick I. Ordway & Randy Lieberman
Washington: Smithsonian Institution Press, 1992
(Study of space theories and notable artworks, developed from a major touring exhibition.)

**THE APOLLO PROGRAM**

***Apollo: The Race to the Moon***
Charles Murray & Catherine Bly Cox
New York: Simon and Schuster, 1989
(Excellent and entertaining account of the lunar landing effort, the building of Apollo and Saturn, plus the creation of NASA's working methods.)

***Powering Apollo: A Biography of James Webb***
W. Henry Lambright
Baltimore: Johns Hopkins University Press, 1995
(An analysis of NASA from Webb's perspective.)

**THE SKYLAB SPACE STATION**

***A House In Space***
Henry S.F. Cooper Jnr.
New York: Holt Rinehart & Winston, 1977
(Entertaining account of life on Skylab, originally a series of articles for the *New Yorker* magazine.)

**THE *CHALLENGER* ACCIDENT**

***Prescription for Disaster***
Joseph P. Trento
New York: Crown, 1987
(Eye-opening account of the background dramas.)

**SPACELAB**

***Spacelab: An International Success Story***
Douglas B. Lord
Washington: GPO, NASA SP–487, 1987
(Full account of NASA–ESA collaboration.)

**THE 'FREEDOM' SPACE STATION**

***The Space Station: A Personal Journey***
Hans Mark
Durham, NC: Duke University Press, 1987
(Mark's experiences within NASA.)

***Space Station: Policy, Planning and Utilization***
Mireille Gerard & Pamela Edwards
New York: American Institute of Aeronautics & Astronautics, 1983
(Authoritative account of official policies.)

***Keeping the Dream Alive: Managing the Space Station Program, 1982–1986***
Thomas J. Lewin & V.K. Narayanan
NASA Contractor Report 4272, 1990
(Details of Hodge Working Group, etc.)

***Together in Orbit: The Origins of International Participation In Space Station Freedom***
John Logsdon
Washington: Space Policy Institute, George Washington University, 1991
(Invaluable account of the origins of the program.)

**GENERAL REFERENCE**

***The Illustrated Encyclopedia of Space Technology***
Kenneth Gatland
London: Salamander, (revised edition, 1989)
(Full technical details of hardware from 1957 to 1989.)

***NASA: A History of the Civil Space Program***
Roger D. Launius
Malaba: Krieger, 1994
(Overview of NASA's complete history to date.)

***The Heavens and the Earth: A Political History of the Space Age***
Walter A. McDougall
New York: Basic Books, 1985
(The links between space adventure and politics.)

UNITED
STATES